# Current Topics in Microbiology and Immunology

Ergebnisse der Mikrobiologie und Immunitätsforschung

54



*With 28 Figures*

Springer-Verlag Berlin · Heidelberg · New York 1971

ISBN 3-540-05289-5 Springer-Verlag Berlin Heidelberg New York
ISBN 0-387-05289-5 Springer-Verlag New York Heidelberg Berlin

## Table of Contents

## List of Contributors

RUTH ARNON, Department of Chemical Immunology, The Weizmann Institute of Science, Rehovot/Israel

THOMAS J. GILL III, Laboratory of Chemical Pathology, Department of Pathology, Harvard Medical School, 25 Shattuck Street Boston, MA 02115/USA

HANS KÜNTZEL, Max-Planck-Institut für experimentelle Medizin, D-3400 Göttingen, Hermann-Rein-Str. 3

M. SERVIN-MASSIEU, Instituto Politécnico Nacional, Escuela National de Ciencias Biológicas, México 17, D. F./México

D. SULITZEANU, Department of Immunology, Hebrew University — Hadassah Medical School, Jerusalem/Israel

Department of Immunology, Hebrew University — Hadassah Medical School

# Antibody-like Receptors on Immunocompetent Cells

D. SULITZEANU

With 2 Figures

Contents



## 1. Introduction

It is probably not an exaggeration to say that one of the central problems of immunology is: how does a cell recognize an antigen when it sees one? BURNET was probably the first to recognize clearly the importance of the recognition problem and also the first to attempt to offer a solution (BURNET and FENNER, 1949). His original indirect template theory, based on so-called recognition units for self-markers, was too much of an exercise in imagination to gain wide acceptance. When he advanced his next theory, the "clonal selection theory" (BURNET, 1959), it naturally encountered strong opposition by the chemically oriented, and by then firmly entrenched, instructive theories. However, as immunology progressed at an ever increasing rate, more and more facts accumulated, which could be best explained in terms of clonal selection. The word "receptor" began to appear in the literature.

I propose to try and bring together in this review some of the indirect evidence for the existence of receptors, i.e. antibody-like sites on the surface of immunocompetent cells. I will then describe the more direct evidence and will finally discuss some of the problems and inconsistencies that still await solution.

## 2. Indirect Evidence for Receptors

### a) Immunoglobulin-like Molecules on Lymphoid Cells

The first indication that immunoglobulin-like molecules may be present on the surface of lymphoid cells came from the work of GELL and SELL (1965) on blast transformation with anti-allotypic sera. Their experiments showed conclusively that rabbit lymphocytes carry genetically determined allotypic markers identical to those found on immunoglobulin molecules. The markers were present even on cells taken from neonatal animals (SELL and GELL, 1965), i.e. long before the cells became functional as immunoglobulin producers. These investigations, further supported by the work of HERZENBERG on allotype suppression (HERZENBERG et al., 1967), could be taken as fairly strong evidence that immunoglobulins participate as structural components in the formation of the lymphocyte surface. The autoradiographic and fluorescent antibody experiments of RAFF et al. (1970) seem to prove this point beyond any doubt.

### b) Receptors and the Secondary Response

An appropriate system for the study of the receptor is the secondary response, in which recognition plays a central role. A few years ago, FAZEKAS DE ST. GROTH and WEBSTER (1966a and b) published two papers with an original title — Disquisitions on original antigenic sin — in which they re-examined an old observation by FRANCIS (1953) and others: if a man or animal is primed by contact with one antigen (e.g. influenza virus of a certain serotype), boosting with a crossreacting antigen will produce antibodies reacting better with the primary antigen. The authors proposed that priming induces in cells a trapping mechanism for the antigen used in priming. This mechanism can also trap the cross reacting antigen, which will then proceed to stimulate cells to produce antibodies, but of the kind which they had been intended to produce, i.e. directed towards the original antigen. The nature of the trapping mechanism was left open, but the investigators mentioned sessile antibody — in other words receptors — as a likely explanation.

It is well known by now that lymphoid cells taken from primed animals are stimulated to incorporate $H^3$-thymidine *in vitro*, when brought into contact with antigen (DUTTON, 1967). This reaction has the specificity of the antibody response — in other words cells are only stimulated by the homologous antigen — which can only mean that the primed cell must somehow recognize the antigen. The simplest solution is the presence of specific receptors. CROSS and MÄKELÄ (1968) stimulated *in vitro* cells from a mouse primed with NIP (4-OH, 3 I-5-$NO_2$ phenylacetic acid) — chicken globulin and transferred the cells to syngeneic irradiated mice. As expected, the recipients gave a secondary type of response. But when free hapten was added to the incubation mixture, the antibody response was partly inhibited (see also MITICHISON, 1967). The conclusion is inescapable that the hapten had competed with the conjugate for some site with specificity directed to the hapten. A different type of inhibition experiments, carried out by FELDMAN and his colleagues (SEGAL et al., 1969) led to similar results. These workers prepared affinity labelling reagents: N-bromoacetyl ε-N-DNP-lysine (BADL) and N-bromoacetyl-N-DNP-ethylene diamine (BADE). These compounds bind covalently to anti-DNP-antibodies and block their combining sites. The blocking is specific, since it can be inhibited by excess hapten (DNP-lysine). The reagents were able to block stimulation of antibody formation by antigen, when added to spleen cultures of primed mice (see also PLOTZ, 1969).

Further support for the receptor theory has come from the experiments of STEINER and EISEN (1967a, b) and from those of the group led by BENACERRAF (PAUL et al., 1967a and b; PAUL et al., 1968; SISKIND et al., 1968). These experiments were concerned with the influence of a number of factors on the affinity of antibodies for antigen. If a large enough dose of antigen is given, the average affinity of the antibodies produced tends to be low. The explanation is that, under these conditions, all cells capable of responding are stimulated. If a small dose is given, however, only cells bearing high affinity receptors will be able to intercept the antigen and be stimulated. Therefore the overall affinity of the antibodies produced will be high. This finding enables us to understand the well known increase in affinity of antibodies with duration of immunization. With the passing of time, the concentration of antigen remaining in the body decreases, so that only cells with high affinity receptors are stimulated. Therefore, late antibodies bind antigens (including crossreacting antigens — LITTLE and EISEN, 1969) more strongly.

High affinity antibodies are also produced when animals primed with a hapten-protein conjugate are boosted with hapten bound to a different carrier. Apparently, only cells with high affinity receptors are capable of interacting with the heterologous antigen and therefore only these are stimulated. Similar considerations apply in tolerance (PAUL et al., 1967a; THEIS et al., 1969). If animals are made tolerant by administration of high doses of antigen, all immunocompetent members of the clones are affected. But if the dose is insufficient, only high affinity cells will bind antigen and be inactivated. Low affinity cells will remain active and will produce low affinity antibody. Moreover, cells bearing crossreacting receptors may not react sufficiently with the original antigen — say BSA — to become tolerant, but they might react with sufficient affinity with the crossreacting antigen, e.g. DNP-BSA, to produce antibodies. This is how one explains breakdown of tolerance to BSA, due to crossreacting DNP-BSA. PAUL et al. (1967a) have shown that antibodies produced under these conditions have a higher affinity for DNP-BSA than for BSA.

Yet another phenomenon involving receptors is the regulation of the antibody response. A number of investigators (reviewed by UHR, 1968; MÖLLER et al., 1968) have shown in recent years that if animals producing antibody are treated with more antibody of the same specificity, especially antibody of the 7S class, further antibody production is stopped. The mechanism of this regulatory effect is not known with certainty but the general opinion is that excess antibody in the circulation competes for antigen with the antibody-like surface receptors on the competent cells.

### c) Receptors and Tolerance

It has been assumed for some time that antigen reacting directly with the competent cells induces tolerance, whereas "processed antigen" i.e. antigen which has first been taken up by macrophages, induces antibody production (reviewed by SULITZEANU, 1968). This hypothesis, although based on much indirect evidence, could not be tested directly because of the difficulties of inducing tolerance *in vitro*. However this has now become possible. We were able recently to paralyze lymphoid cells of primed mice by exposing them *in vitro* to fairly low doses of antigen (BIRNBAUM and SULITZEANU, 1969). Comparable results were obtained by BYERS and SERCARZ (1968) with primed lymph node fragments and by BRITTON (1969) with normal mouse lymphoid cells. A most convincing demonstration of "*in vitro*" induced tolerance has also been provided by DIENER and ARMSTRONG (1969). These investigators exposed mouse spleen cells *in vitro* either to a small dose (20 ng) of flagellin of *S. waycross* or to a small dose ($4 \times 10^6$) of SRBC and, simultaneously, to a tolerizing dose (1 μg) of *S. adelaide* flagellin. The cells were then transferred to lethally irradiated recipients, which were subsequently challenged with the same pair of antigens. The recipients responded poorly to the tolerizing antigen but gave a normal response to the other antigen. Tolerance was produced even when the cells were exposed to the antigen pair at 4°. The inescapable conclusion from this and the other experiments is that the large antigen dose paralyzed the corresponding clone, most likely by reacting with receptors on the cell surface.

## 3. Precommitment

It would seem appropriate to interrupt for a moment this account of factual evidence in order to emphasize one point which may not be immediately evident. To suscribe to the idea of receptor is tantamount to accepting the idea of precommitment. If lymphoid cells are precommitted, they must be able to recognize an antigen without having ever seen it before. And to recognize the antigen they, and only they, must possess the instrument of recognition — the specific receptor. The experiments of DIENER and ARMSTRONG (1969) are a nice illustration of this point. Another illustration, obtained with a different but most interesting experimental system, is provided by the work of ABDOU and RICHTER (1969a). Rabbits were irradiated and their competence restored by injecting them with allogeneic bone marrow cells. Rabbits thus treated give a good antibody response to sheep red blood cells (SRBC). However, if the bone marrow donors are injected with SRBC 24 hours before the cell transfer, the recipients are incapable of responding to an injection of SRBC, although their response to a different RBC (horse) remains unimpaired. The experiment shows again that only one type of clone is somehow inactivated in this system — the clone(s) capable of interacting with SRBC in the donor.

WIGZELL and ANDERSON (1969) used an "*in vitro*" device to deplete a specifically reacting cell clone. They transferred to irradiated recipients primed cells mixed with antigen (HSA, BSA or OA), in order to stimulate an adoptive secondary response. When the lymphoid cells were first passed through columns of glass or plastic beads coated with antigen, the immune cells were selectively retained by the column and the eluted cells lost the ability to transfer memory. Retention could be blocked specifically by free antigen. To show that retention was not due to cytophilic antibody, the investigators passed cells from animals immunized to two antigens through columns containing only one of the antigens. As expected, only the right kind of response was lost under these circumstances. These results entitled WIGZELL and ANDERSON to conclude that the primed lymphoid cells were retained on the column through surface cell receptors capable of reacting with the antigen. ABDOU and RICHTER (1969b) went one important step further. Not only did they succeed in depleting a *normal* cell population (as contrasted to the primed cells of WIGZELL and ANDERSON) of antigen reacting cells (ARCs) — they also managed to recover the cells from the column, obtaining, in effect, what seems to be a population of cells enriched in ARCs. Obviously, if one could do this routinely with reasonable yields, the way would be open for the isolation of cell receptors.

## 4. Direct Evidence for Receptors

I have discussed thus far experiments which can be interpreted most easily in terms of receptors. I will now describe another line of evidence, which might be taken to constitute more direct proof for the existence of receptors. I am referring here to a line of research initiated in our laboratory some four years ago, which has since been taken up by ADA and his group and by HUMPHREY and KELLER. It occurred to us quite a number of years ago that, if BURNET'S theory is correct, it should be possible to detect the presence of antibody-like sites on the lymphocyte membrane by reacting the lymphocytes directly with a highly labelled antigen. Obviously, the labelled antigen should bind to lymphocytes carrying the receptor and to them alone. This idea was so simple and straightforward that, in retrospect, it is surprising that so few people had thought of trying it. The explanation is perhaps, that although the word receptor was coming into print at an ever increasing rate, people did not take it quite seriously.

When this work was started, we thought it necessary to avoid two major pitfalls. First, non-specific binding and second, binding of antigen due to pinocytosis. To minimize non specific binding, an antigen (BSA) was used with a high specific activity (100 $\mu$Ci/$\mu$g) since in this way one could work with very little protein in the binding test (0.01 $\mu$g or less). To minimize uptake due to pinocytosis, the reaction was performed in the cold (4°). The results of these experiments were quite interesting. A small number of the normal spleen and lymph node cells became labelled after contact with antigen in the cold (NAOR and SULITZEANU, 1967; SULITZEANU and NAOR, 1969). One cell in 1,500 was highly labelled, but since about half of the labelled cells were macrophages, it would be more correct to say that one out of 3,000 lymphoid cells was highly labelled (i.e. contained over 1,600 molecules of antigen). These figures are

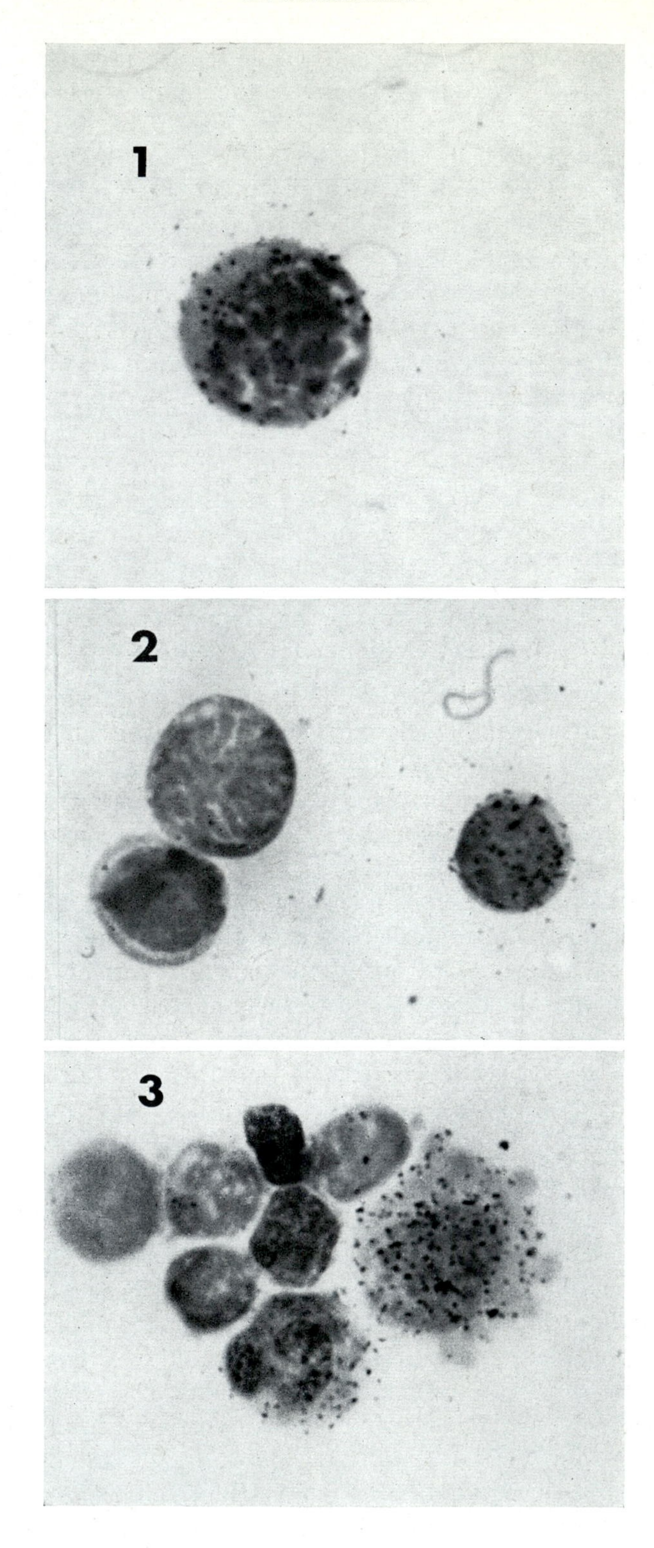


Fig. 1

astonishingly similar to the figure quoted by BIOZZI (BIOZZI et al., 1968) for rosette forming cells (RFC) in the spleen of normal mice, which is 1 : 1,500. The labelled cells (Fig. 1) did not belong to a single morphological class. On the contrary, many cell types were seen, ranging from medium lymphocytes to large blast cells. Quite unexpectedly, however, typical small lymphocytes, of the size of RBC, were very rarely labelled. When similar binding experiments were done with cells taken from immunized mice, a much higher proportion of the cells became labelled and the intensity of labelling of the individual cells was higher (NAOR and SULITZEANU, 1969a). Many macrophages were also strongly labelled, since they were probably carrying cell-bound cytophilic antibody. The expectation that cells taken from tolerant donors would show reduced binding appeared to be confirmed, since this is what we found in three replicate trials (NAOR and SULITZEANU, 1969b). Unfortunately, one deals in these experiments with such low numbers of labelled cells that the significance of the differences observed must remain to a large extent doubtful.

The interpretation of these results was by no means simple. One could be certain that the data relating to cells from immunized donors were dependable, because of the large numbers of labelled cells recovered. On the other hand, the significance of the results obtained with normal cells was uncertain. Were we dealing here with an artifact, totally unrelated to the point we were trying to make? One experiment led us to believe that this was unlikely. When normal mouse lymphoid cells were exposed to a mixture of antigens, actually four iodinated albumins (Table page 8), the number of labelled cells was much higher than when the experiment was carried out with one albumin only (SULITZEANU and NAOR, 1969). We concluded, therefore, that the albumins were bound to specific sites, present on distinct cell clones.

A very similar type of study was performed by BYRT and ADA (1969), but they carried it much further. BYRT and ADA tested the binding of flagellar proteins and of hemocyanin to mouse and rat cells, in a reaction system which included sodium azide, to prevent phagocytosis of antigen by macrophages. They labelled the antigens with $I^{131}$, which produces so much blackening that the labelled cells become visible at low magnification. This enabled them to scan considerably more cells than we could — $1—5 \times 10^5$ cells. In so far as their experiments paralleled ours, there were few important differences. Perhaps the

Fig. 1. Membrane-bound receptors for bovine serum albumin (BSA) on mouse lymphoid cells. The cells were exposed to highly iodinated $^{125}$I-BSA at 4°, washed, smeared on slides and subjected to autoradiography. Each silver grain represents approximately 100 molecules. **1**) Blast-like cell from lymph node of normal mouse. [SULITZEANU and NAOR: The affinity of radioiodinated BSA for lymphoid cells. II. Binding of $^{125}$I-BSA to lymphoid cells of normal mice. Int. Arch. Allergy **35**, 564—578 (1969).] **2**) Medium lymphocyte from spleen of mouse immunized to BSA. [NAOR and SULITZEANU: Affinity of radioiodinated Bovine Serum Albumin for lymphoid cells; Binding of $^{125}$I-BSA to lymphoid cells of immune mice. Israel J. med. Sci. **5**, 217—229 (1969).] **3**) Peritoneal cells from mouse immunized to BSA. The macrophages carry cytophilic antibody to BSA, by means of which the labelled antigen becomes attached to the macrophage cell membrane

Table. *Binding of four radioiodinated albumins to mouse spleen cells*

| Antigen used in binding experiments | No of cells with grain counts | | | | | Total cells surveyed |
|---|---|---|---|---|---|---|
| | 0—3 | 4—6 | 7—10 | 11—15 | 16 or more | |
| Mixture of antigens | 5,667 | 238 | 62 | 20 | 13 | 6,000 |
| BSA | 5,843 | 115 | 29 | 6 | 7 | 6,000 |
| DSA | 5,968 | 25 | 4 | 2 | 1 | 6,000 |
| GPSA | 5,858 | 34 | 4 | 1 | 3 | 6,000 |
| HSA | 5,987 | 9 | 3 | - | 1 | 6,000 |

Spleen cells of normal mice were exposed at 4° to a mixture of 4 radioiodinated albumins (Bovine, Dog, Guinea Pig and Human) or to each of these albumins separately. The total amount of radioactivity used for each binding test was $4 \times 10^6$ cpm. This was divided equally among the 4 antigens in the tube containing the antigen mixture. The distribution of labelled cells was examined in autoradiographs. The proportion of labelled cells in the cell suspension exposed to the mixture of antigens was much higher than in any of the cell suspensions exposed to a single antigen, indicating that different albumins were bound to different cell clones (from SULITZEANU and NAOR). The affinity of radioiodinated BSA for lymphoid cells. II. Binding of 125-I-BSA to lymphoid cells of normal mice. Int. Arch. Allergy 35: 564—578, 1969 (S. Karger, Basel/New York).

major one concerns the type of the labelled lymphocytes, which were in their opinion of the small variety. The outstanding contribution of ADA's work was the demonstration that the binding of antigen to normal lymphoid cells, far from being an artifact, has indeed an immunological significance. To prove this, lymphoid cells were exposed to flagellin in the usual way, kept at 0° *in vitro* to allow the labelled cells to be damaged by irradiation and then transferred to syngeneic, irradiated animals. One day later, the recipients were challenged with an immunogenic dose of the flagellin used in the binding tests and with a similiar dose of an unrelated flagellin. The results of this beautiful experiment (ADA and BYRT, 1969) were clear cut: antibody was produced only against the unrelated antigen, which must mean that the cells which had bound the labelled flagellin *in vitro* were the cells that would have otherwise responded to this flagellin.

Binding experiments were also performed by HUMPHREY and KELLER (1969) with the synthetic, multichain polypeptide TIGAL [(T, G)-A--L 509; SELA et al., 1962] and with haemocyanin and their results are quite similar to those already mentioned. The antigens they used had an extremely high specific activity — 1,400 Ci/g, so that practically each molecule could be visualized (20—30 grains per molecule in autoradiography). Yet the distribution of labelled cells was, in general, not different from that described by us or ADA.

HUMPHREY and KELLER tried to induce primary and secondary antibody responses with the labelled antigens. With the very highly labelled preparations, they got neither a primary response, nor priming and not even secondary responses. In contrast, TIGAL prepared with non-labelled iodide had all these activities. As in the case of ADA's work, the only explanation must be in terms of a specific radiation damage inflicted on precommitted, receptor-bearing cells.

## 5. Receptors on Cells Mediating Delayed Hypersensitivity

Compared to the compelling evidence for the existence of receptors on antibody-producing cells, there is very little support for the existence of receptors involved in delayed hypersensitivity. DAVID and SCHLOSSMAN (1968) were able to inhibit the migration of peritoneal cells of guinea pigs sensitized to α-DNP-polylysine, by adding the corresponding antigen to the cultures. Only the heptamer or the higher peptides were effective. This requirement for the higher peptides in order to elicit the *in vitro* reaction is paralleled by a similar requirement for the elicitation of the reaction *in vivo*. This experiment may certainly be taken to demonstrate specific recognition. ROITT and his colleagues (GREAVES et al., 1969) found recently that anti-light chain serum, or its Fab fragment, could suppress the mitogenic response to tuberculin, as well as the mixed lymphocyte reaction, implying that the receptors involved in these reactions may have an immunoglobulin-like structure. One need hardly mention the obvious objection that the effect of such treatments could be simply due to steric hindrance. DAGUILLARD and RICHTER (1969) showed that rabbit thymus cells treated with goat anti-rabbit IgG serum failed to give a blastogenic response, although the cells did respond to cellular mitogenic agents (allogeneic or xenogeneic cells). The authors concluded on this basis that thymus cells can only mediate cellular immunity and that such cells have no surface recognition sites. It would seem far too early, however, to accept this view, even if the evidence on which it was based were much stronger. The fact remains that it is extremely difficult, if not altogether impossible, to conceive of any immunological reactivity not requiring stereospecific interactions and it is equally difficult to conceive of such interactions taking place at any other location except at the cell surface.

## 6. Receptors for the Carrier Molecule

At the same time that the receptor theory was becoming respectable, it was becoming clear that the immunocyte receptor alone was not adequate to account for all the phenomena related to the induction process (JERNE, 1967). Since the receptor was supposed to possess a specificity more or less resembling that of the antibody, it could only interact with the determinant group. However, a large variety of experimental data indicated that other parts of the antigen molecule played a role in the induction process — the so called carrier effect (MITICHISON, 1967; review: PLESCIA, 1969). The first indication for the carrier effect came from experiments on the elicitation of secondary responses to hapten — protein conjugates. As a rule, secondary responses to the hapten can be elicited only by injecting the hapten conjugated to the original protein carrier; this in spite of the fact that much of the antibody produced is specific for the hapten alone. A carrier effect has also been found in the induction of tolerance and in delayed hypersensitivity. Probably the most striking demonstration of the carrier effect is provided by the work of RAJEWSKI et al. (1967) on the immune response to lactic dehydrogenase (LDH) isozymes.

The LDH system comprises several enzymes, built of 4 polypeptide subunits, of the type AAAA, BBBB or AABB. If rabbits are primed with AABB

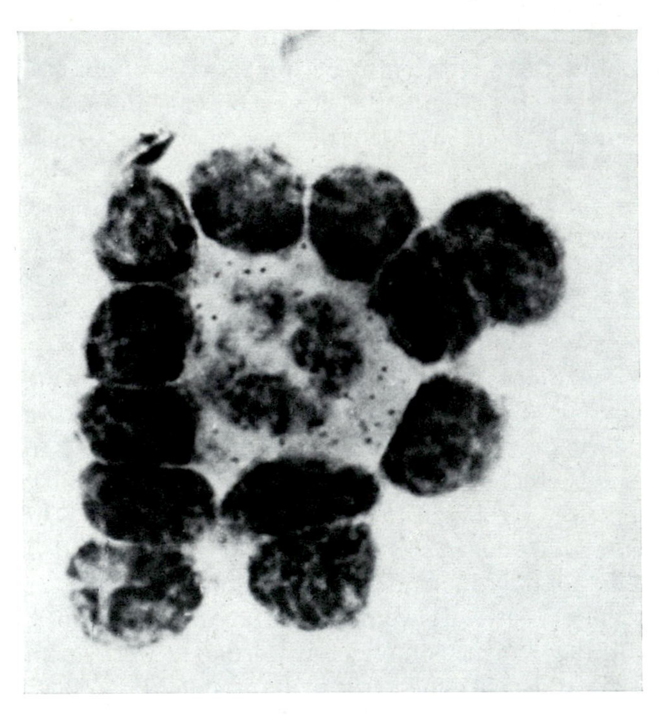

Fig. 2. Presumed mechanism of the carrier effect. Peripheral lymphoid cells of rabbit sensitized to BSA, cultured for 3 days in the presence of 1 $\mu$g $^{125}$I-BSA. A cluster is seen, consisting of a central macrophage surrounded by lymphocytes. The macrophage has taken up the labelled BSA, thus serving as a "carrier" cell for the carrier portion of the BSA molecule. Any lymphoid cell in the cluster capable of responding to a BSA determinant (i.e. a cell belonging to the BSA "clone"), would be stimulated to produce antibody to BSA [SULITZEANU, KLEINMAN, BENEZRA, and GERI: Nature (Lond.) (in press)]

and boosted with the same antigen, antibodies are produced, as one would expect, which react both with A and with B. Therefore AABB can prime to both antigens. If AAAA is used as booster, the animal responds by producing A antibodies. But if BBBB is used as booster, the response is weak or absent. This means that, although the animals are primed to B, they cannot respond to B alone, but must be stimulated with AB. A plays the role of the carrier. The carrier effect has led MITICHISON (1967) to postulate the existence of a second type of receptor capable of recognizing the carrier and of interacting with it. The well-known role of cellular cooperation in the immune response (MILLER and MITCHELL, 1969; MOSIER and COPPLESON, 1968) suggests that the carrier receptor might be found on a different cell type, possibly on the memory cell (ROITT et al., 1969), on the "mediator" cell postulated by PLESCIA (1969), or on the macrophage. It is more than likely that the antigen-reactive cells (ARC) in the normal animal and the memory cells (produced in response to the carrier determinants) in the sensitized animal, are in effect the "mediator cells" of PLESCIA. This view is strongly supported by the work of RAJEWSKI et al. (1969): rabbits primarily stimulated with the BSA-sulfanilic acid conjugate gave a secondary response to HGG-sulfanilic acid only if also immunized to HGG. It stands to reason that the memory cells, produced by the administration of HGG, provided the receptors on which the HGG-sulfanilic acid molecules could become anchored, in order to stimulate another cell type (antibody forming cells — AFC) to produce antibodies to the hapten. This explanation makes it immediately apparent why blast cell-lymphocytes clusters should form in antigen stimulated cultures of primed lymphocytes (SULITZEANU et al., in press). The central cell in such clusters (Fig. 2) probably binds the carrier portion of the antigen molecule, with the haptenic determinants free to stimulate cells of the appropriate clone, should any of them be among the peripheral cells in the cluster. The fact that macrophage-lymphocytes clusters are also found in high frequency in such cultures suggests that macrophages might also act as carriers for antigen, either non specifically or by means of cytophilic antibodies functioning as "specific" receptors.

So far, no direct evidence has been found for true antigen receptors on the macrophages. A unique nucleic acid in rat macrophages has been shown by GOTTLIEB (1969) to complex with antigen taken up by these cells, but there is little to implicate this material as a *bona fide* receptor. On the other hand, there are good indications that macrophages have distinct receptors for immunoglobulin molecules and therefore for antibodies, of both the 7S and 19S classes (LAY and NUSSENZWEIG, 1969; HUBER et al., 1969; HENSON, 1969). The antibodies, on their part (in particular the cytophilic antibodies — reviewed by SULITZEANU, 1968) may act for all intents and purposes as indirect macrophage receptors for antigen. The role, if any, of such interactions between antigens and macrophage-bound cytophilic antibodies in the induction process remains yet to be uncovered. Additional receptors have been postulated, specific for the complement components (LAY and NUSSENZWEIG, 1968, 1969). It seems in fact that, once the inhibitions against receptors will have fallen, we are likely to be deluged by hordes of them and separating the genuine from the imaginary is going to become a problem.

## 7. Characteristics of Receptors

Purists will raise an eyebrow at the thought of someone trying to describe an entity, the very existence of which has not been yet definitely proven. Nonetheless, it may not be entirely worthless to summarize the little we know, or think we know, about what receptors are like. The receptors of the non-stimulated lymphoid cells are, perhaps, IgM-like molecules (DWYER and MACKAY, 1970), distributed on the cell surface in a patchy fashion (ADA, personal communication). We do not know the immunoglobulin type of the receptors present on stimulated cells but it would be certainly interesting to see whether they are IgG like. Receptors seem to appear early in life, since they have been found on cells from a 22 week old human fetus (DWYER and MACKAY, 1970). As mentioned before, there is a correlation between the immunological state and the number of cells bearing receptors. There are more such cells in immune animals, and, possibly, fewer in tolerant animals. In a limited experiment with human peripheral blood cells, lymphocytes of agammaglobulinemic patients displayed a lower affinity for foreign antigens than lymphocytes of normal persons, indicating a reduced number of receptor-bearing cells (NAOR et al., 1969).

Most experiments suggest that cells are generally precommitted to one antigen (PLAYFAIR et al., 1965; PLAYFAIR, 1968; PAPERMASTER, 1967; SULITZEANU and NAOR, 1969; ADA and BYRT, 1969; WIGZELL and ANDERSON, 1969; ABDOU and RICHTER, 1969), in other words, that they carry receptors of a single specificity. To the examples I gave so far I will add OSOBA's experiments, which are quite clear cut. OSOBA (1969) worked with MARBROOK's *in vitro* system (1967), with limiting dilutions of spleen cells stimulated simultaneously with two species of RBC — sheep and chicken. He found some cultures containing PFC for SRBC and others containing PFC for CRBC. Evidently, these cells could recognize and react to one type of RBC only. The specificity of the receptor is generally similar to that of the antibody the cell is destined to produce. Thus, excess hapten can block stimulation of secondary responses by the conjugate and this block can again be reversed by excess hapten-protein conjugate (MITICHISON, 1967). However, the specificity of the receptor might be higher than that of the antibodies. As an example, guinea pig cells sensitized to α-DNP-PLL by an immunogenic member of the series (heptamer or larger) can be stimulated to incorporate thymidine only by an immunogenic member of the series (STULBERG and SCHLOSSMAN, 1968; SCHLOSSMAN et al., 1969). The receptor has, therefore, specificity for the heptamer or for a higher member, although the antibodies produced on challenge with the heptamer will react quite well with the hapten alone. Their specificity is less stringent. It is not unlikely that receptors are also specialized to conform with the specialization of the immunoglobulin-producing cell, e.g. cells producing IgM or IgG, PFC or RFC (SHEARER et al., 1968; SHEARER and CUDKOVICZ, 1969; SHEARER et al., 1969). The affinity of the receptor reflects the affinity of the antibody the cell will produce, as shown by the changes in antibody affinity mentioned in section 2b. Since we know so little about the receptors, most questions remain open. It would be important to know, for instance, whether the receptors are structural components of the cell membrane or whether they are merely normal antibodies in the process of being secreted by the cell.

## 8. Comments

It is clear that to accept the receptor one has to accept the basic tenet of the clonal theory in practically its original form. Many people will find this difficult at a time when some experimental work is still incompatible with the theory. I am referring here to the experiments of ADLER et al. (1966) and NISBET et al. (1969), to the evidence on antigenic competition (reviewed by ADLER, 1964) and to recent findings on the excessive frequency of PFC (NOSSAL et al., in press). These phenomena, while stressing the complexity of the problem, cannot be considered as arguments strong enough to refute the receptor concept.

A complication has recently arisen from the newly discovered specialization of immunocompetent cells into cells reacting with antigen (ARC) and those producing antibody (MILLER and MITCHELL, 1969). Do both cell types have specific receptors? There appears to be little doubt that the ARCs do. It has been shown (ABDOU and RICHTER, 1969c) that the immune response of rabbits tolerant to HSA or BGG can be restored with normal bone marrow cells, which are the source of ARCs in the rabbit. This cannot be done with cells from donors tolerant to the same antigen as the recipient. In mice, in which ARCs are derived from the thymus (MILLER and MITCHELL, 1969), tolerance has been shown repeatedly to involve this organ (TAYLOR, 1968; ARGYRIS, 1968; ARMSTRONG et al., 1969; see also ISAKOVIĆ et al., 1965, and STAPLES et al., 1966, for experiments with rats). It would seem therefore, that the cells affected in the tolerant state are the ARCs and the specificity of tolerance can be explained in terms of ARCs bearing specific receptors. PLAYFAIR (1969), on the other hand, found that in mice the specificity of the antibodies is determined by the marrow cells, which are the source of antibody-forming cells. While, as admitted by PLAYFAIR (1969), it is possible that both cell types have antigen specific receptors, it is perplexing to note that mouse thymus cells have no demonstrable affinity for antigen in the radioactive antigen binding tests (BYRT and ADA, 1969; HUMPHREY and KELLER, in press; NAOR and SULITZEANU, in press), whereas the bone marrow cells bind the label in large numbers, probably non-specifically.

Once the receptor theory is accepted, the prolonged discussion as to whether competent cells are uni- or multipotential must be considered as settled. Evidently, it would be hard to assume the existence of an infinite range of receptors on the individual cell membrane and experiments such as those of ADA and RICHTER strongly support the idea of unipotent cells. Also, the confusion surrounding the terms precommitted and committed cells is resolved, since these terms become identical. All committed cells are in fact precommitted. The receptor theory also requires that the antigen, whatever its tribulations in the body, should finally react with the receptor of the immunocompetent cell, whether to induce a primary or secondary response or to induce tolerance. To demonstrate that this is so one has to show the presence of antigen in the stimulated, immunocompetent cell. NOSSAL has found indeed antigen in antibody-forming cells (NOSSAL et al., 1967), but his *in vivo* experiments could not be interpreted unequivocally. We have been pursuing the same problem using a system more likely to give a dependable answer (Birnbaum and Sulitzeanu, unpublished experiments). Lymphoid cells of mice primed to BSA were incubated with $I^{125}$-labelled BSA *in vitro,* washed to remove excess antigen and placed in diffusion chambers. Most of the labelled cells placed in the chambers were lymphocytes, with practically no blast cells. Three days later, however, there were no labelled lymphocytes in the diffusion chambers, but a fair number of labelled blasts. It is reasonable to assume that some of the labelled blasts were derived from memory cells transformed as a result of direct interaction with the antigen they contained.

One does not need an exceptional prophetic talent or an inordinate amount of optimism to predict that receptors will be soon isolated, analyzed and catalogued. This, after all, would only be a minor problem compared to the next major challenge — to discover the link between receptor activation and mechanics of cell stimulation. The model proposed by Bretcher and Cohn (1968), which postulates that antigen induces conformational changes in the receptor molecule, is interesting, but, to say the least, incomplete, as it disregards completely the role of the ARC. A fair guess would be that, following attachment of antigen to receptor, a major imbalance in the normal metabolism of the cell membrane might occur. There are now reasons to believe that the cell membrane, far from being a static structure, is undergoing constant turnover (Warren and Glick, 1968). It is not difficult to imagine that, with receptors frozen by bound antigen, the turnover might be seriously impaired. How this is translated into cell division and immunoglobulin production remains a formidable problem, but the interesting analogy, recently noted by Braun and his coworkers (Ishizuka et al., in press) between hormone-mediated and antigen-mediated cellular activation, might point the way to an eventual solution.

*Acknowledgments*

Some of the work from our laboratory cited in this review was carried out with the aid of grants from the Concern Foundation of Los Angeles, the Sam Lautenberg Fellowship, and the Bogen Fund.

## References

Abdou, N. I., Richter, M.: Cells involved in the immune response. VI. The immune response to red blood cells in irradiated rabbits after administration of normal, primed, or immune allogeneic rabbit bone marrow cells. J. exp. Med. **129**, 757—774 (1969a).

— — Cells involved in the immune response. X. The transfer of antibody-forming capacity to irradiated rabbits by antigen-reactive cells isolated from normal allogeneic rabbit bone marrow after passage through antigen-sensitized glass bead columns. J. exp. Med. **130**, 141—163 (1969b).

— — Cells involved in the immune response. XI. Identification of the antigen-reactive cell as the tolerant cell in the immunologically tolerant rabbit. J. exp. Med. **130**, 165—184 (1969c).

ADA, G. L., BYRT, P.: Specific inactivation of antigen-reactive cells with $^{125}$I-labelled antigen. Nature (Lond.) **222**, 1291—1292 (1969).
ADLER, F. L.: Competition of antigens. Progr. Allergy **8**, 41—57 (1964).
— FISHMAN, M., DRAY, S.: Antibody formation initiated in vitro. III. Antibody formation and allotypic specificity directed by ribonucleic acid from peritoneal exudate cells. J. Immunol. **97**, 554—558 (1966).
ARGYRIS, B. F.: Role of thymus in adoptive tolerance. J. Immunol. **100**, 1255—1258 (1968).
ARMSTRONG, W. D., DIENER, E., SHELLAM, G. R.: Antigen-reactive cells in normal, immunized, and tolerant mice. J. exp. Med. **129**, 393—410 (1969).
BIOZZI, G., STIFFEL, C., MOUTON, D., BOUTHILLIER, Y., DECREUSEFOND, C.: A kinetic study of antibody producing cells in the spleen of mice immunized intravenously with sheep erythrocytes. Immunology **14**, 7—20 (1968).
BIRNBAUM, M., SULITZEANU, D.: Immunological paralysis of sensitized cells, induced in vitro. Immunology **17**, 635—637 (1969).
BRETSCHER, P. A., COHN, M.: Minimal model for the mechanism of antibody induction and paralysis by antigen. Nature (Lond.) **220**, 444—448 (1968).
BRITTON, S.: Regulation of antibody synthesis against Escherichia coli endotoxin. IV. Induction of paralysis in vitro by treating normal lymphoid cells with antigen. J. exp. Med. **129**, 469—481 (1969).
BURNET, F. M., FENNER, F.: Production of antibodies, 2nd ed. Melbourne: Macmillan 1949.
BURNET, M.: The clonal selection theory of acquired immunity. Cambridge: University Press 1959.
BYERS, V. S., SERCARZ, E. E.: The X-Y-Z scheme of immunocyte maturation. V. Paralysis of memory cells. J. exp. Med. **128**, 715—728 (1968).
BYRT, P., ADA, G. L.: An in vitro reaction between labelled flagellin or haemocyanin and lymphocyte-like cells from normal animals. Immunology **17**, 503—516 (1969).
CROSS, A. M., MÄKELÄ, O.: Selective inhibition of the secondary response of primed cells by incubation with hapten. Immunology **15**, 389—394 (1968).
DAGUILLARD, F., RICHTER, M.: Cells involved in the immune response. XII. The differing responses of normal rabbit lymphoid cells to phytohemagglutinin, goat anti-rabbit immunoglobulin antiserum and allogeneic and xenogeneic lymphocytes. J. exp. Med. **130**, 1187—1208 (1969).
DAVID, J. R., SCHLOSSMAN, S. F.: Immunochemical studies on the specificity of cellular hypersensitivity. The in vitro inhibition of peritoneal exudate cell migration by chemically defined antigens. J. exp. Med. **128**, 1451—1459 (1968).
DIENER, E., ARMSTRONG, W. D.: Immunological tolerance in vitro: Kinetic studies at the cellular level. J. exp. Med. **129**, 591—603 (1969).
DUTTON, R. W.: In vitro studies of immunological responses of lymphoid cells. Advan. Immunol. **6**, 253—336 (1967).
DWYER, J. M., MACKAY, I. R.: Antigen-binding lymphocytes in human blood. Lancet **1970 I**, 164—167.
FAZEKAS DE ST. GROTH, S., WEBSTER, R. G.: Disquisitions on original antigenic sin. I. Evidence in man. J. exp. Med. **124**, 331—345 (1966a).
— — Disquisitions on original antigenic sin. II. Proof in lower creatures. J. exp. Med. **124**, 347—361 (1966b).
FRANCIS, T.: Influenza: The Newe Acquayantance: Ann. intern. Med. **39**, 203—221 (1953).
GELL, P. G. H., SELL, S.: Studies on rabbit lymphocytes in vitro. II. Induction of blast transformation with antisera to six IgG allotypes and summation with mixtures of antisera to different allotypes. J. exp. Med. **122**, 813—821 (1965).
GOTTLIEB, A. A.: Studies on the binding of soluble antigens to a unique ribonucleoprotein fraction of macrophage cells. Biochemistry **8**, 2111—2116 (1969).

GREAVES, M. F., TORRIGIANI, G., ROITT, I. M.: Blocking of the lymphocyte receptor site for cell mediated hypersensitivity and transplantation reactions by anti light chain sera. Nature (Lond.) **222**, 885—886 (1969).
HENSON, P. M.: The adherence of leucocytes and platelets induced by fixed IgG antibody or complement. Immunology **16**, 107—121 (1969).
HERZENBERG, L. A., HERZENBERG, L. A., GOODLIN, R. C., RIVERA, E. C.: Immunoglobulin synthesis in mice. Suppression by anti-allotype antibody. J. exp. Med. **126**, 701—713 (1967).
HUBER, H., MICHLMAYR, G., FUDENBERG, H. H.: The effect of anti-lymphocyte globulin on human monocytes in vitro. Clin. exp. Immunol. **5**, 607—617 (1969).
HUMPHREY, J. H., KELLER, H. U.: Some evidence for specific interaction between immunologically competent cells and antigens. In: Developmental aspects of antibody formation and structure (J. STERZL and I. RIHA, eds.). London-New York: Academic Press (in press).
ISAKOVIĆ, K., SMITH, S. B., WAKSMAN, B. H.: Role of the thymus in tolerance. I. Tolerance to bovine gamma globulin in thymectomized, irradiated rats grafted with thymus from tolerant donors. J. exp. Med. **122**, 1103—1123 (1965).
JERNE, N. K.: Summary: Waiting for the end. In: Cold Spr. Harb. Symp. quant. Biol. **32**, 591—603 (1967).
LAY, W. H., NUSSENZWEIG, V.: Receptors for complement on leukocytes. J. exp. Med. **128**, 991—1007 (1968).
— — $Ca^{++}$ dependent binding of antigen — 19 S antibody complexes to macrophages. J. Immunol. **102**, 1172—1178 (1969).
LITTLE, J. R., EISEN, H. N.: Specificity of the immune response to the 2,4-dinitrophenyl and 2,4,6-trinitrophenyl groups. Ligand binding and fluorescence properties of crossreacting antibodies. J. exp. Med. **129**, 247—265 (1969).
MARBROOK, J.: Primary immune response in cultures of spleen cells. Lancet **1967 II**, 1279—1281.
MILLER, J. F. A. P., MITCHELL, G. F.: Thymus and antigen reactive cells. In: Transplan. Reviews (G. MÖLLER, ed.) **1**, 3—42 (1969).
MITCHISON, N. A.: Antigen recognition responsible for the induction *in vitro* of the secondary response. In: Cold Spr. Harb. Symp. quant. Biol. **32**, 431—439 (1967).
MÖLLER, E., BRITTON, S., MÖLLER, G.: Homeostatic mechanisms in cellular antibody synthesis and cell-mediated immune reactions. In: Regulation of the antibody response (B. CINADER, ed.), p. 141—181. Springfield, Ill.: Ch. C. Thomas 1968.
MOSIER, D. F., COPPLESON, L. W.: A three-cell interaction required for the induction of the prim ry immune response *in vitro*. Proc. nat. Acad. Sci. (Wash.) **61**, 542—547 (1968).
NAOR, D., BENTWICH, Z., CIVIDALLI, G.: Inability of peripheral lymphoid cells of agammaglobulinaemic patients to bind radioiodinated albumins. Aust. J. exp. Biol. med. Sci. **47**, 759—761 (1969).
— SULITZEANU, D.: Binding of radioiodinated bovine serum albumin to mouse spleen cells. Nature (Lond.) **214**, 687—688 (1967).
— — Affinity of radioiodinated bovine serum albumin for lymphoid cells. Binding of $I^{125}$-BSA to lymphoid cells of immune mice. Israel J. med. Sci. **5**, 217—229 (1969a).
— — Binding of $^{125}I$-BSA to lymphoid cells of tolerant mice. Int. Arch. Allergy **36**, 112—113 (1969b).
— — Affinity of radioiodinated bovine serum albumin for lymphoid cells. III. Further experiments with cells of normal animals. Israel J. Med. Sci. (in press).
NISBET, N. W., SIMONSEN, M., ZALESKI, M.: The frequency of antigen-sensitive cells in tissue transplantation. A commentary on clonal selection. J. exp. Med. **129**, 459—467 (1969).

NOSSAL, G. J. V., BUSSARD, A. E., LEWIS, H., MAZIE, J. C.: Formation of hemolytic plaques by peritoneal cells in vitro. I. A new technique enabling micromanipulation and yielding higher plaque numbers. In: Developmental aspects of antibody formation and structure (J. STERZL and I. RIHA, eds.). New York: Academic Press (in press).
— WILLIAMS, G. M., AUSTIN, C. M.: Antigens in immunity. XIII. The antigen content of single antibody forming cells early in primary and secondary immune responses. Aust. J. exp. Biol. med. Sci. **45**, 581—594 (1967).
OSOBA, D.: Restriction of the capacity to respond to two antigens by single precursors of antibody-producing cells in culture. J. exp. Med. **129**, 141—152 (1969).
PAPERMASTER, B. W.: The clonal differentiation of antibody-producing cells. Cold Spr. Harb. Symp. quant. Biol. **32**, 447—460 (1967).
PAUL, W. E., SISKIND, G. W., BENACERRAF, B.: A study of the "termination" of tolerance to BSA with DNP-BSA in rabbits: relative affinities of the antibodies for the immunizing and the paralyzing antigens. Immunology **13**, 147—157 (1967a).
— — — Specificity of cellular immune responses. Antigen concentration dependance of stimulation of DNA synthesis in vitro by specifically sensitized cells, as an expression of the binding characteristics of cellular antibody. J. exp. Med. **127**, 25—42 (1968).
— — — OVARY, Z.: Secondary antibody responses in haptenic systems: cell population selection by antigen. J. Immunol. **99**, 760—770 (1967b).
PLAYFAIR, J. H. L.: Strain differences in the immune response of mice. II. Responses by neonatal cells in irradiated adult hosts. Immunology **15**, 815—826 (1968).
— Specific tolerance to sheep erythrocytes in mouse bone marrow cells. Nature (Lond.) **222**, 882—883 (1969).
— PAPERMASTER, B. W., COLE, L. J.: Focal antibody production by transferred spleen cells in irradiated mice. Science **149**, 998—1000 (1965).
PLESCIA, O. J.: The role of the carrier in antibody formation. In: Current topics in microbiology and immunology, vol. **50**, p. 78—106. Berlin-Heidelberg-New York: Springer 1969.
PLOTZ, P. H.: Specific inhibition of an antibody response by affinity labelling. Nature (Lond.) **223**, 1373—1374 (1969).
RAFF, M. C., STERNBERG, M., TAYLOR, R. B.: Immunoglobulin determinants on the surface of mouse lymphoid cells. Nature (Lond.) **225**, 553—554 (1970).
RAJEWSKY, K., ROTTLÄNDER, E., PELTRE, G., MÜLLER, B.: The immune response to a hybrid protein molecule. Specificity of secondary stimulation and of tolerance induction. J. exp. Med. **126**, 581—606 (1967).
— SCHIRRMACHER, V., NASE, S., JERNE, N. K.: The requirement of more than one antigenic determinant for immunogenicity. J. exp. Med. **129**, 1131—1143 (1969).
ROITT, I. M., GREAVES, M. F., TORRIGIANI, G., BROSTOFF, J., PLAYFAIR, J. H. L.: The cellular basis of immunological responses. A synthesis of some current views. Lancet **1969 II**, 367—371.
SCHLOSSMAN, S. F., HERMAN, J., YARON, A.: Antigen recognition: in vitro studies on the specificity of the cellular immune response. J. exp. Med. **130**, 1031—1045 (1969).
SEGAL, S., GLOBERSON, A., FELDMAN, M., HAIMOVICH, J., GIVOL, D.: Specific blocking of antibody synthesis *in vitro* by affinity labelling reagents. Nature (Lond.) **223**, 1374—1375 (1969).
SELA, M., FUCHS, S., ARNON, R.: Studies on the chemical basis of the antigenicity of proteins. Synthesis, characterization and immunogenicity of some multichain and linear polypeptides containing tyrosine. Biochem. J. **85**, 223—235 (1962).
SELL, S., GELL, P. G. H.: Studies on rabbit lymphocytes in vitro. IV. Blast transformation of the lymphocytes from newborn rabbits induced by anti-allotype serum to a paternal IgG allotype not present in the serum of the lymphocyte donors. J. exp. Med. **122**, 923—928 (1965).

SHEARER, G. M., CUDKOWICZ, G.: Cellular differentiation of the immune system of mice. III. Separate antigen-sensitive units for different types of anti sheep immunocytes formed by marrow-thymus cell mixtures. J. exp. Med. **129**, 935—951 (1969).

— — CONNELL, M. ST. JAMES, PRIORE, R. L.: Cellular differentiation of the immune system of mice. I. Separate splenic antigen sensitive units for different types of anti sheep antibody-forming cells. J. exp. Med. **128**, 437—457 (1968).

— — PRIORE, R. L.: Cellular differentiation of the immune system of mice. II. Frequency of unipotent splenic antigen-sensitive units after immunization with sheep erythrocytes. J. exp. Med. **129**, 185—199 (1969).

SISKIND, G. W., DUNN, P., WALKER, J. G.: Studies on the control of antibody synthesis. II. The effect of antigen dose and of suppression by passive antibody on the affinity of antibody synthesized. J. exp. Med. **127**, 55—66 (1968).

STAPLES, P. J., GERY, I., WAKSMAN, B. H.: Role of the thymus in tolerance. III. Tolerance to bovine gamma globulin after direct injection of antigen into the shielded thymus of irradiated rats. J. exp. Med. **124**, 127—139 (1966).

STEINER, L. A., EISEN, H. N.: Sequential changes in the relative affinity of antibodies synthesized during the immune response. J. exp. Med. **126**, 1161—1183 (1967a).

— — The relative affinity of antibodies synthesized in the secondary response. J. exp. Med. **126**, 1185—1205 (1967b).

STULBERG, M., SCHLOSSMAN, S. F.: The specificity of antigen-induced thymidine-2-$^{14}$C incorporation into lymph node cells from sensitized animals. J. Immunol. **101**, 764—769 (1968).

SULITZEANU, D.: Affinity of antigen for white cells and its relation to the induction of antibody formation. Bact. Rev. **32**, 404—424 (1968).

— KLEINMAN, R., BENEZRA, D., GERY, I.: Cellular interactions and the secondary response in vitro. Nature (Lond.) (in press).

— NAOR, D.: The affinity of radioiodinated BSA for lymphoid cells. II. Binding of $^{125}$I-BSA to lymphoid cells of normal mice. Int. Arch. Allergy **35**, 564—578 (1969).

TAYLOR, R. B.: Immune paralysis of thymus cells by Bovine Serum Albumin Nature (Lond.) **220**, 611 (1968).

THEIS, G. A., GREEN, I., BENACERRAF, B., SISKIND, G. W.: A study of immunologic tolerance in the dinitrophenyl poly-L-Lysine immune system. J. Immunol. **102**, 513—518 (1969).

UHR, J. W.: Inhibition of antibody formation by serum antibody. In: Regulation of the antibody response (B. CINADER, ed.), p. 114—126. Springfield, Ill.: Ch. C. Thomas 1968.

WARREN, L., GLICK, M. C.: The metabolic turnover of the surface membrane of the L cell. In: Biological properties of the mammalian surface membrane. The Wistar Institute Symposium monograph No 8 (L. A. MANSON, ed.), p. 3—15. Philadelphia: The Wistar Institute Press 1968.

WIGZELL, H., ANDERSON, B.: Cell separation on antigen-coated columns. Elimination of high rate antibody-forming cells and immunological memory cells. J. exp. Med. **129**, 23—36 (1969).

Laboratory of Chemical Pathology
Department of Pathology
Harvard Medical School, Boston, MA 02115

# Synthetic Polypeptide Metabolism[1]

THOMAS J. GILL III[2]

With 8 Figures

Contents



## Introduction

The chemistry and the biological fate of the antigen are two major parameters in determining whether the antigen interacts with the immunocompetent cell. Interest in the metabolic fate of synthetic polypeptides developed from attempts to explain the differences in immunogenicity between polypeptides composed of L-amino acids and those composed of D-amino acids. The major areas of interest have been the degradation and organ localization of the antigens and the correlation of the nature and magnitude of the antibody response with the fate of the antigen.

The metabolism of protein antigens has been investigated extensively in a variety of species in order to explore the induction of the immune response. These studies focused either on correlating the persistence of antigen with the appearance of antibody (CAMPBELL and GARVEY, 1963, 1965; RICHTER et al., 1965) or on localizing antigen in tissues, presumably at its site of action (COHEN et al., 1966; NOSSAL et al., 1968a, 1968b; MILLER et al., 1968). However, experiments designed to determine explicitly the site of antigen action may face the same sorts of difficulties as might be encountered in trying to study fertilization by investigating the blastula: detailed inspection of the

1 The nomenclature of the synthetic polypeptides is modified from that defined in the Tentative Rules on Abbreviated Nomenclature of Synthetic Polypeptides [Europ. J. Biochem. 3, 129—131 (1967)]. The other abbreviations are defined in their context. Ab is the symbol for antibody throughout.

2 The author is the recipient of a Research Career Development Award from the National Institutes of Health (K3-AM-5242). The research carried out in his laboratory was supported by grants from the National Science Foundation (GB-8379) and from the National Heart Institute (HE-1771).

developing embryo would not reveal the nature of the initiating event. At the current stage of knowledge about the cellular basis of the antibody response, it may be equally as difficult to discern the mechanism of antigen action.

## Studies in Rabbits

Early studies on the catabolism of poly($Glu^{56}Lys^{38}Tyr^{6}$) (GILL and DAMMIN, 1962) showed that the $^{131}$I-polypeptide was rapidly eliminated from the circulation following intravenous injection. The elimination pattern did not vary significantly among the primary, sensitized and anamnestic responses, if the persistence of free iodine or small iodine-containing peptides in the serum was taken into account (GILL et al., 1964b; GILL et al., 1965); this was a problem only with large doses, e.g., 60 mg. Passage through Sephadex G-25 equilibrated with 0.11 M NaCl-0.04 M phosphate buffer, pH 6.8, separated polypeptide-bound isotope from free iodine or iodine attached to small peptides (CARPENTER et al., 1967). The average serum volume of rabbits was measured as $4.4 \pm 0.3$ % of the body weight, and this value was used in all calculations (GILL et al., 1964b).

The method of iodination did not affect the elimination of the labeled polypeptide from the serum, the excretion of degradation products in the urine, or the localization of the polypeptide in organs (GILL et al., 1965; CARPENTER et al., 1967). Both the iodine monochloride method (MCFARLANE, 1958) and the method using free iodine in alkaline solution (TALMAGE et al., 1954) gave the same results. Enzymatic hydrolysis and chromatographic analysis of $^{131}$I-labeled poly($Glu^{58}Lys^{36}Tyr^{6}$) showed that the iodine label was present as 48% mono-iodo-tyrosine, 22% di-iodo-tyrosine, and 7% thyroxin; the rest was found either as free iodine or remained at the origin.

Studies of the role of polypeptide metabolism in immunogenicity became more cogent following the observation that D-amino acid polymers were not nearly as immunogenic as their L-amino acid counterparts. Using moderate amounts of antigen for immunization (20 mg), 80 μg Ab/ml were elicited by poly($Glu^{58}Lys^{42}$) (6/12 animals responding), but no antibody was elicited by poly($_DGlu^{57}{}_DLys^{43}$) (0/24 animals) (GILL et al., 1963). The observation that poly($_DGlu^{57}{}_DLys^{43}$) did not elicit antibody formation was ascribed either to a failure in the sequence of steps leading to antibody production or to an inability of the $\gamma$-globulin chain to fold around the determinant portion of the D-polypeptide. The latter explanation was soon discounted on theoretical grounds and by the demonstration that poly($_DGlu^{55}{}_DLys^{39}{}_DTyr^{6}$) could elicit an average of 160 μg Ab/ml (13/17 animals) in comparison with the 517 μg Ab/ml elicited in all animals (38/38) immunized with the optical enantiomorph poly($Glu^{56}Lys^{38}Tyr^{6}$) (GILL et al., 1964a). Therefore, it appeared as if there was a defect in the inductive phase of the antibody response: either the D-polypeptide could not be transported to the site of antibody production or some prerequisite hydrolysis could not be performed by the usual enzymes (GILL et al., 1963).

An investigation of the first possibility focused on studying the state in which synthetic polypeptides circulated by vertical and horizontal starch gel electrophoresis of the $^{131}$I-polypeptides in serum (PAPERMASTER et al., 1965). Following electrophoresis, the gels were fixed and stained to localize the protein components of the serum, and then radioautography was done to localize the polypeptides. The in vitro experiments were carried out by mixing 0.4 to 1.1 mg of synthetic polypeptide per ml of normal rabbit, guinea pig, ox or dog serum to give a final specific activity of 10 $\mu$C/mg polymer. All of the sera bound the synthetic polypeptides, and the patterns with the sera from the different species were similar. In vivo studies were performed by injecting 10 mg of synthetic polymer into New Zealand white rabbits and drawing blood samples for electrophoresis at 5 and 20 minutes; the results were the same as those of the in vitro studies. The most extensive studies, which employed rabbit serum analyzed by vertical starch gel electrophoresis at pH 8.6, are summarized in Table 1. The electrophoretic bands of a synthetic polypeptide mixed with serum were generally different from those of the polypeptide itself; therefore, most of the polymer apparently interacted with various serum proteins. The negatively charged polymers bound to more serum proteins than the positively charged ones, and the polymer containing equal amounts of glutamic acid and lysine, poly ($Glu^{47}Lys^{47}Tyr^{6}$), showed binding characteristics intermediate between those of glutamic acid-rich and lysine-rich polymers. The optical configuration of the amino acids did not influence the binding pattern, since poly ($Glu^{56}Lys^{38}Tyr^{6}$) and poly ($_{D}Glu^{55}{}_{D}Lys^{39}{}_{D}Tyr^{6}$) bound in the same way to the serum proteins. Additional electrophoretic experiments using horizontal starch gel electrophoresis in acetate buffer pH 5.5, borate buffer pH 8.6, phosphate buffers pH 6.3 and 7.8, and carbonate buffer pH 10.6 and experiments using the vertical technique in phosphate buffer pH 8.6 all showed that the synthetic polypeptides bound to serum proteins. These studies confirmed the generalization that glutamic acid-rich polymers bound to more serum proteins and that the configuration of the amino acid residues did not affect binding.

There is immunochemical evidence that the binding of synthetic antigens to serum proteins is not necessary for immunogenicity or for antigenic specificity. Fractionated antiserum containing only purified antibody quantitatively precipitated synthetic polypeptides; hence, the antibody was directed against the synthetic antigen alone and not against an antigen-serum protein complex. Also, there was no evidence that any antibody was formed against serum proteins. Radioautographic studies showed that synthetic polypeptides and homologous serum proteins localized to quite different parts of the kidney (CARPENTER et al., 1967); therefore, the binding did not affect their independent catabolism.

Thus, there is no evidence for any specific in vivo coupling of synthetic polypeptide antigens to serum proteins or for the involvement of a specific transport mechanism in the induction of an antibody response to the polypeptides. The range of the antibody responses elicited by the various poly-

Table 1. *Physical chemical properties, immunogenicity and migration on starch gel electrophoresis of a variety of synthetic polypeptide antigens*[a]

| Polypeptide | Net charge | Molecular weight | Per-cent helix | Antibody response | | Number of responders | Number and type of bands in the binding pattern in various regions of serum proteins[b] | | | |
|---|---|---|---|---|---|---|---|---|---|---|
| | | | | μg Ab/ml | ±S.D. | | γ-globulin (origin) | α-globulin | β-globulin | albumin |
| poly($Glu^{11}Lys^{85}Tyr^{4}$) | +74 | 148,000 | 0 | 755 | 718 | 9/9 | 1 broad | 1 sharp | 1 sharp | |
| poly($Glu^{84}Lys^{10}Tyr^{6}$) | −74 | 110,000 | 0 | 350 | 456 | 8/8 | 1 broad | 4 sharp | 1 sharp | 2 sharp |
| poly($Glu^{37}Lys^{58}Tyr^{5}$) | +21 | | 10 | 693 | 362 | 6/6 | 1 broad | | 1 sharp | |
| poly($Glu^{56}Lys^{38}Tyr^{6}$) | −18 | 110,000 | 15 | 569 | 325 | 38/38 | 1 broad | 3 sharp | 2 sharp | 1 broad |
| poly(DGlu$^{55}$DLys$^{39}$DTyr$^{6}$) | −16 | 93,000 | 5 | 162 | 206 | 13/17 | 1 broad | 3 sharp | 2 sharp | 1 broad |
| poly($Glu^{47}Lys^{47}Tyr^{6}$) | 0 | | 50 | 506 | 281 | 7/8 | | 1 sharp | 1 sharp | 1 sharp |
| poly($Lys^{96}Tyr^{4}$) | +96 | | 0 | 38 | | 4/10 | | | 1 sharp | |
| poly($Glu^{96}Tyr^{4}$) | −96 | 114,000 | 0 | 31 | 31 | 5/8 | 2 broad | 1 sharp | 1 sharp | 1 sharp |

[a] PAPERMASTER et al. (1965); GILL et al. (1967).

[b] Vertical starch gel electrophoresis of the polypeptide in serum. The buffer was glycine-NaOH, pH 8.6. The binding patterns were essentially the same whether the antigen was mixed with serum in vitro or injected into a rabbit.

peptides in the glutamic acid-rich group (30 to 570 μg Ab/ml), in the lysine-rich group (40 to 750 μg Ab/ml) and by the polypeptide containing equal amounts of glutamic acid and lysine (500 μg Ab/ml) was the same. No correlation exists between the immunogenicity of a synthetic polypeptide and its serum binding pattern (Table 1).

The second possibility to explain the differences in the immunogenicity of the L- and D-polypeptides is that the catabolism and organ localization of the polypeptides is the crucial factor. Following intravenous injection of 10 mg of

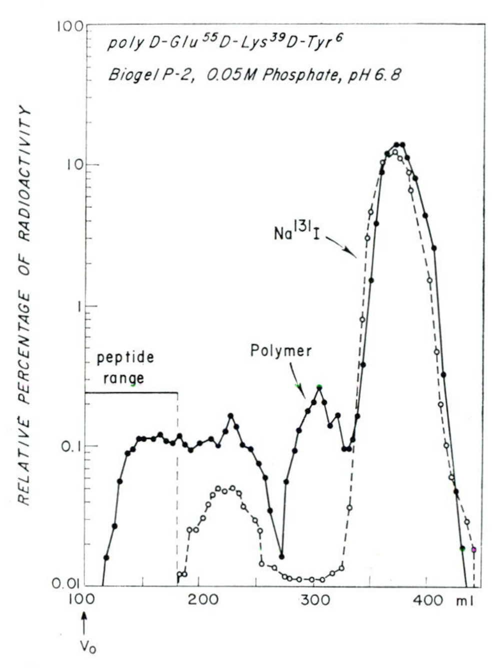


Fig. 1. Urinary excretion of peptides from the degradation of poly ($_D$Glu$^{55}$$_D$Lys$^{39}$$_D$Tyr$^{6}$). The chromatographic analysis of the dialyzable urinary radioactivity obtained from animals injected with the D-polymer is shown by the black circles: a small but consistant amount of material appears in the peptide range. Animals injected with Na$^{131}$I alone are shown for comparison (open circles). (Carpenter et al., 1967)

poly(Glu$^{58}$Lys$^{36}$Tyr$^{6}$) (No. 2) or of poly($_D$Glu$^{55}$$_D$Lys$^{39}$$_D$Tyr$^{6}$), the polypeptide was rapidly eliminated from the serum (Gill et al., 1965). The L-polymer was rapidly degraded and excreted by the fourth day, whereas the D-polypeptide was very slowly degraded, and only 30 to 35% was excreted in three to four weeks. Both the L- and D-polypeptides were equally susceptible to phagocytosis, since 15 to 25% of a 200 μg aliquot of poly(Glu$^{58}$Lys$^{36}$Tyr$^{6}$) or of poly($_D$Glu$^{55}$$_D$Lys$^{39}$$_D$Tyr$^{6}$) could be taken up in 30 minutes by $2 \times 10^6$ macrophages (Gill et al., 1964b). The L-polymer elicited $356 \pm 212$ μg Ab/ml in 15/15 rabbits, and the D-polymer elicited $162 \pm 206$ μg Ab/ml in 13/17 rabbits. The serum elimination rate and degradation of the L-polypeptide were independent of the amount injected, but the amount of the D-polymer degraded was a constant fraction of the amount injected. The radioactivity in the urine was analyzed on a calibrated Biogel P-2 column, and a significant portion of this radioactivity was contained in peptides of molecular weight 1,000 to 1,500 (Fig. 1).

The effect of injecting one isomer on the subsequent metabolism of the other isomer was studied in order to determine whether there was any common metabolic pathway that could be pre-empted by the previous injection of one of the enantiomorphs. In a group of rabbits which were injected with 10 mg

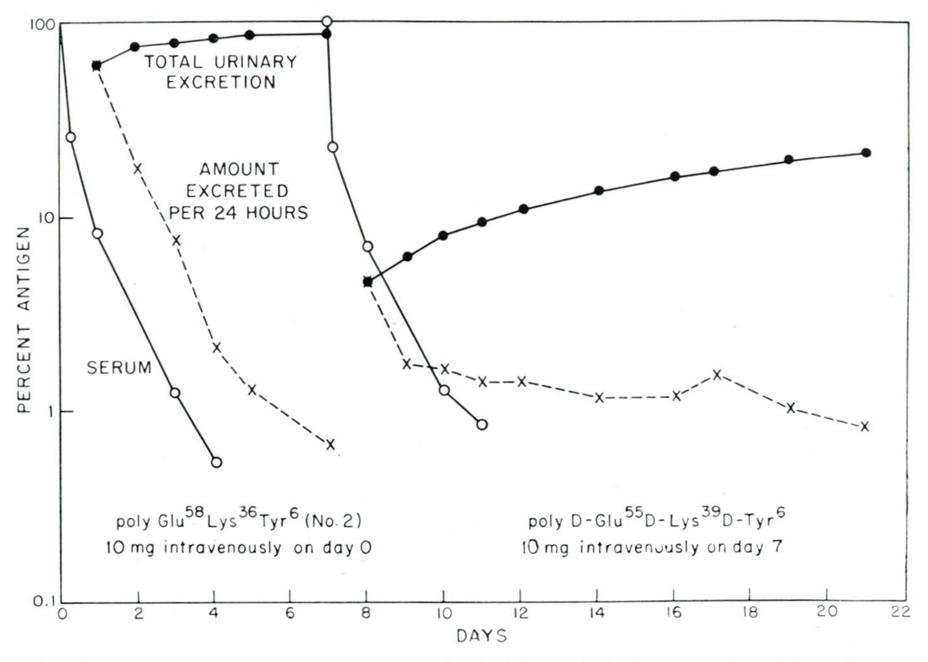


Fig. 2. The effect of injecting 10 mg of poly (Glu$^{58}$Lys$^{36}$Tyr$^{6}$) (No. 2) on the subsequent metabolism of 10 mg of poly (DGlu$^{55}$DLys$^{39}$DTyr$^{6}$) injected 7 days later. The data for the serum elimination and urinary excretion of both polymers are the averages for nine rabbits. (GILL et al., 1965)

of poly(Glu$^{58}$Lys$^{36}$Tyr$^{6}$) (No. 2) on day 0 and subsequently with 10 mg of poly(DGlu$^{55}$DLys$^{39}$DTyr$^{6}$) on day 7, there was no effect on the rate of elimination of the D-polymer from the serum or on the amount of D-polymer degraded and excreted in the urine (Fig. 2). Essentially the same experiment was carried out in the reverse order and again pre-treatment did not cause any effect: the serum elimination pattern and degradation of the L-polymer were the same (Fig. 3). A third experiment was done in which a priming dose of 1 mg of the D-polymer was given two weeks before the 10 mg dose of the D-polymer to see whether the metabolism of the latter amount could be enhanced by priming and then to test the effects of both of these treatments on the subsequent metabolism of the L-polymer. Priming had no effect on the degradation of the second dose of D-polymer, and the two doses of the D-

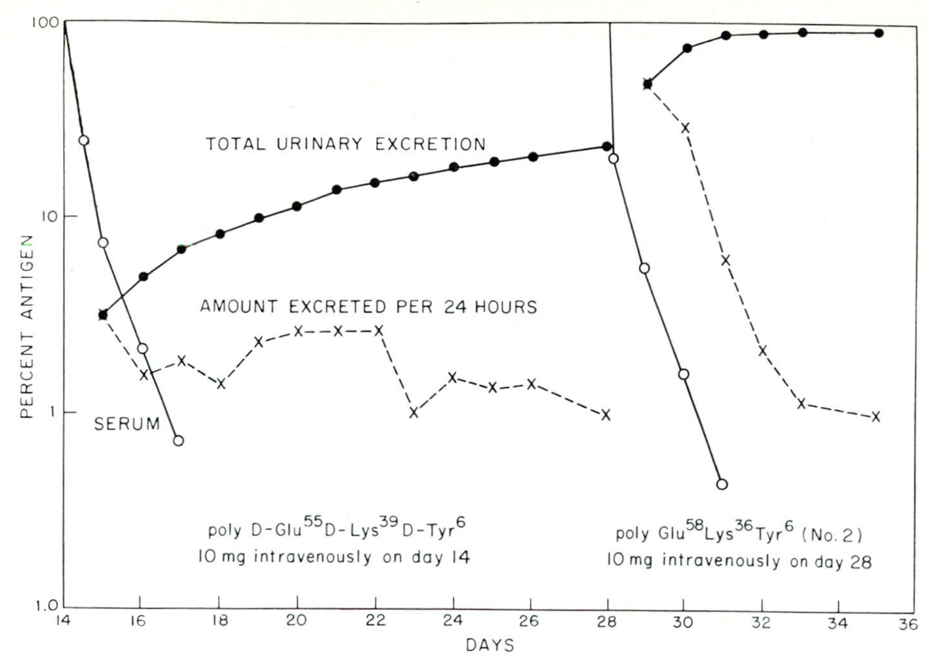


Fig. 3. The effect of injecting 10 mg of poly(DGlu$^{55}$DLys$^{39}$DTyr$^{6}$) on the subsequent metabolism of 10 mg of poly(Glu$^{58}$Lys$^{36}$Tyr$^{6}$) (No. 2). The data are the averages for four rabbits. (GILL et al., 1965)

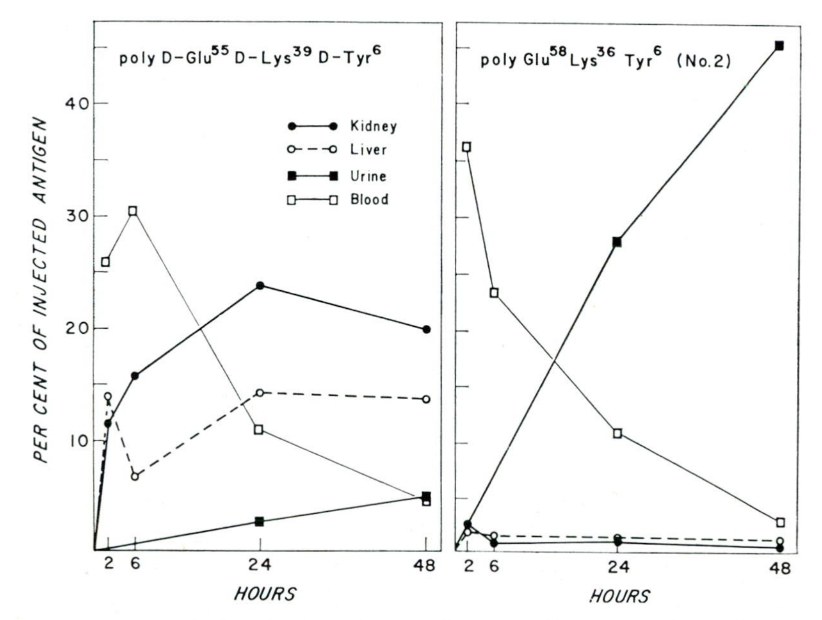


Fig. 4. Degradation and distribution of isomeric synthetic polypeptides. Ten milligrams of poly(DGlu$^{55}$DLys$^{39}$DTyr$^{6}$) or poly(Glu$^{58}$Lys$^{36}$Tyr$^{6}$) (No. 2) labeled with $^{131}$I were injected on day 0. The percentage of injected antigen is the percentage of non-dialyzable radioactivity in blood, liver and kidneys. The serum elimination rates for the polymers are similar, but the D-polymer is retained in the organs to a much greater extent. (CARPENTER et al., 1967)

polymer did not have any effect on the metabolism of the L-polymer. Thus, it appears that the enzymes involved in its metabolism are not induced by the D-polymer and that there are different pathways of degradation for the isomeric synthetic polypeptides.

The difference in metabolism of the D- and the L-polymers is paralleled by differences in the organ retention of the two polypeptides (Carpenter et al., 1967). The fates of the D- and L-isomers during the first 48 hours following injection are shown in Fig. 4. The serum elimination rates were similar, but the L-polymer was degraded at least ten-fold faster. The maximal

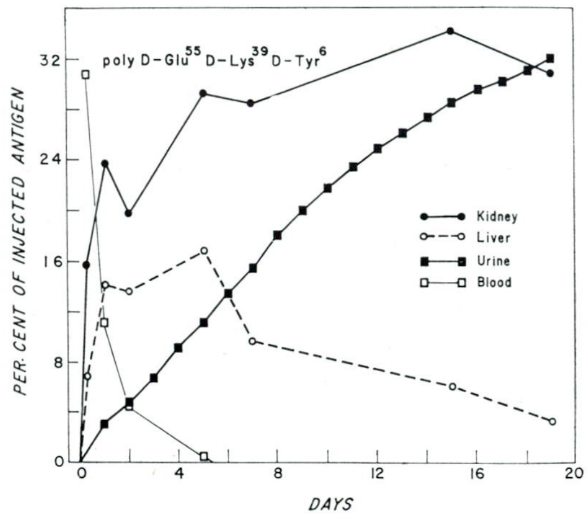


Fig. 5. Degradation and distribution of poly(DGlu$^{55}$DLys$^{39}$DTyr$^{6}$) over a 19-day period (non-dialyzable radioactivity). The urinary excretion rate of $^{131}$I is about 2% of the injected dose/day. The amount of polymer in the liver declines slowly after the first week, whereas the amount in the kidney continues to rise during the second and third weeks. (Carpenter et al., 1967)

amounts of L-polymer in the liver and kidney were less than 2.5% of the injected dose, but both of these organs contained 10 to 15% of the injected dose of D-polymer within the first two hours. The amounts of L-polymer in the liver and kidney declined after two hours, but those of the D-polymer continued to rise (Fig. 5); other organs contained small quantities of D-polymer. The L-polymer was virtually unmeasurable in any tissue after three to five days (Fig. 4 and Table 2). The amount of D-polymer in the liver declined after the first week, but that in the kidney continued to rise, reaching a maximum of 30 to 35% of the injected dose in three to four weeks (Fig. 5). The retained material contained a large fraction of peptide-bound radioactivity, which represented intact antigen or large fragments thereof (Fig. 6). The slow decline in the amount of D-polymer in the liver and the small but steady urinary excretion of radioactivity indicated that the liver was an important site of polypeptide degradation. Comparison of the specific activities of liver at 24 and 48 hours with the specific activities of plasma in the same animal showed that there was an eightfold increase in dialyzable radioactivity in liver compared to plasma for animals injected with the D-polymer, but no such gradient existed in animals injected with the L-polymer. This observation provides further evidence that in the liver of animals injected with the D-polymer there were peptides which were not in free diffusion equilibrium with the extracellular fluid.

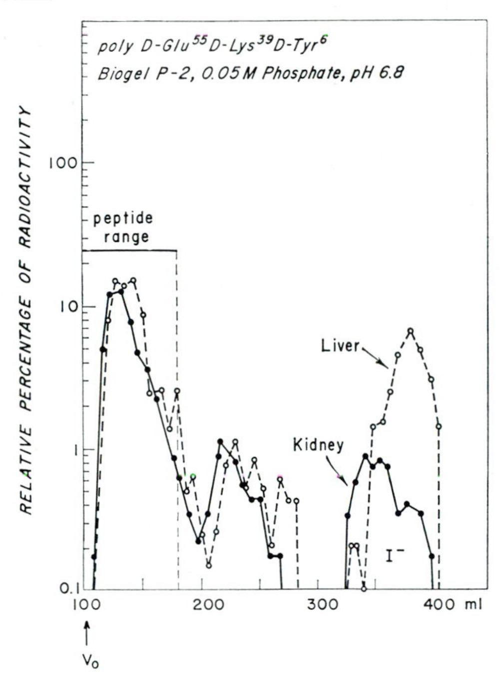


Fig. 6. Organ localization of peptides from the degradation of poly(DGlu$^{55}$DLys$^{39}$DTyr$^{6}$). The curves show the distribution of dialyzable radioactivity obtained from homogenates of liver and kidney obtained 24 hrs after injection of the D-polymer. A large portion of the radioactivity from both organs is contained in peptides, and the patterns are virtually identical for the liver and the kidney. (CARPENTER et al., 1967)

Table 2. *Retention of enantiomorphic polypeptides in various organs after intravenous injection*[a]

| Polypeptide | Day | Liver | Kidney | Spleen | Lungs | Intestine | Adrenals |
|---|---|---|---|---|---|---|---|
| poly (Glu$^{58}$Lys$^{36}$Tyr$^{6}$) | 1 | 1 | 0.8 | 0.1 | 0.1 | 0.6 | 0 |
| (No. 2) | 7 | 0.2 | 0.1 | 0 | 0 | 0 | 0 |
| poly (DGlu$^{55}$DLys$^{39}$DTyr$^{6}$) | 1 | 14 | 24 | 1 | 0.3 | 1.0 | 0.02 |
| | 7 | 10 | 29 | 0.7 | 0.2 | 0.5 | 0.05 |
| | 19[b] | 3 | 31 | 0.6 | 0.1 | 0.4 | 0.04 |

[a] GILL et al. (1965); CARPENTER et al. (1967). Ten mg of polymer labeled with $^{131}$I by the iodine monochloride method were used in each case.

[b] None in lymph nodes, thymus, muscle.

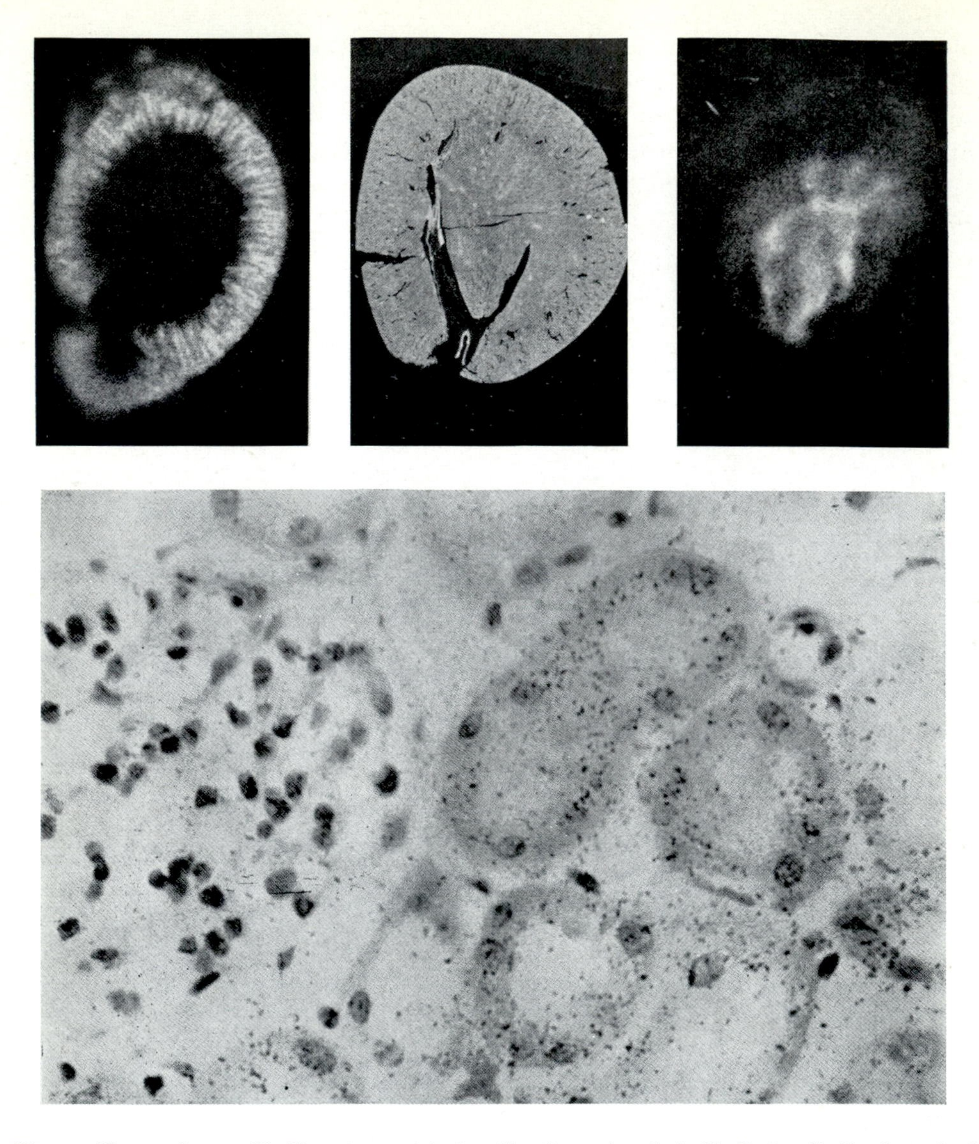

Fig. 7. *Upper frame.* Radioautographic localization of poly(DGlu$^{55}$DLys$^{39}$DTyr$^{6}$) and of autologous serum proteins. The center photograph shows the anatomic areas of the rabbit kidney: there are distinct boundaries between the cortical area, which contains tubular cells and most of the glomeruli, the inner and outer medullary regions, and the papilla. The radioautograph on the left shows the cortical localization of the polymer; 8 mg of polymer were injected intravenously and the whole kidney had a specific activity of 475 cpm/mg. The radioautograph on the right shows the medullary localization of autologous serum proteins. The specific activity of the injected proteins was the same as that for the polymer, but following the intravenous injection of 8 mg of protein, the specific activity of the kidney was only 11 cpm/mg. This finding indicates that only very small amounts of autologous serum proteins remain in the kidney, and they are in the medulla. *Lower frame.* Radioautograph of the cortical region of the kidney from an animal injected with $^{131}$I-labeled poly(DGlu$^{55}$DLys$^{39}$DTyr$^{6}$) 19 days previously. The radioactivity is present in the cytoplasm of the proximal tubular cells. The scattered grains over the glomeruli are no greater than background. (CARPENTER et al., 1967)

Radioautographic studies showed that the retained D-polymer was localized in the cortex of the kidney, where it was in the proximal tubular cells (Fig. 7); the pattern of localization was the same throughout the entire three-week period following injection of the polymer. The L-polymer showed an identical localization during the first few hours after injection, although much less material was present. Autologous serum proteins (8—10 mg) labeled to the same specific activity as the polymers localized in the medullary zone in very small amounts (Fig. 7); thus, the renal localization of the polypeptides was not dependent upon their interaction with plasma proteins. Since there is adequate evidence that the proximal tubular cells participate in the normal catabolism of proteins (HUGHES, 1956; OLIVER and MACDOWELL, 1958; SOLOMON et al., 1964), the persistence of the D-polymer indicates that the enzymes necessary for its degradation are lacking in the kidney. However, there was no histological evidence of renal damage caused by the D-polymer, nor was there any evidence for the combination of antibody or complement with the retained D-polymer (CARPENTER et al., 1967).

Table 3. *The localization of poly(DGlu$^{55}$DLys$^{39}$DTyr$^{6}$) and of autologous serum proteins in kidneys with patent and with ligated ureters*[a]

| Kidney | Percentage of injected polypeptide or protein | | | |
|---|---|---|---|---|
| | poly(DGlu$^{55}$DLys$^{39}$DTyr$^{6}$) | | Autologous serum proteins | |
| | 20 min | 2 hrs | 20 min | 2 hrs |
| Patent ureter | 2.6 | 8.1 | 0.1 | 0.2 |
| Ligated ureter | 1.1 | 2.4 | 0.2 | 0.1 |
| Localization | Cortex (proximal tubular cells) | | Medulla | |

[a] CARPENTER, et al. (1967). The polypeptide and the serum proteins were labeled with $^{131}$I by the iodine monochloride method.

In order to investigate how the polypeptides entered the renal tubules, studies were done with acute ureteral ligation. Acute ligation of one kidney, with the other being used as the control, caused a striking reduction in the amount of D-polymer entering the proximal tubular cells (Table 3). This finding indicates that most of the polypeptide was normally filtered by the glomerulus and then absorbed. In contrast, there was no significant effect of ureteral ligation on the uptake of serum proteins, since they are normally not filtered in appreciable quantities by the glomerulus. However, minute amounts may enter the tubular cells from the peritubular capillaries (SHUSTER et al., 1963).

The importance of the kidney in the metabolism of the D-polypeptide was further emphasized by studies in anephric animals (Table 4): the serum level of the polypeptide remained higher than in rabbits with kidneys, but localization

Table 4. *The effect of nephrectomy on the retention of poly(*D*Glu*$^{55}$D*Lys*$^{39}$D*Tyr*$^{6}$*) in the serum and organs*[a]

| Organ | Percentage of poly(DGlu$^{55}$DLys$^{39}$DTyr$^{6}$) retained | | | | | | Percentage of poly(Glu$^{58}$Lys$^{36}$Tyr$^{6}$) (No. 2) retained | | |
|---|---|---|---|---|---|---|---|---|---|
| | Anephric rabbits[b] | | | Normal rabbits | | | Normal rabbits | | |
| | 6 hrs | 24 hrs | 48 hrs | 6 hrs | 24 hrs | 48 hrs | 6 hrs | 24 hrs | 48 hrs |
| Serum | 35 | 24 | 7 | 26 | 10 | 4 | 23 | 11 | 3 |
| Liver | 8 | 9 | 16 | 7 | 14 | 14 | 2 | 1 | 0.8 |
| Kidney | — | — | — | 16 | 24 | 20 | 0.9 | 0.8 | 0.4 |
| Spleen | 1 | 1 | 2 | 0.5 | 1 | 1 | 0.1 | 0.1 | 0.1 |
| Lungs | 0.2 | 0.1 | 0.2 | 0.5 | 0.3 | 0.3 | 0.1 | 0.1 | 0.1 |
| Intestine | 0.5 | 0.7 | 2 | | 1.0 | | 0.8 | 0.6 | 0.7 |
| Adrenals | <0.1 | <0.1 | <0.1 | | <0.1 | | 0 | 0 | 0 |

[a] CARPENTER et al. (1967). The polypeptides were labeled with $^{131}$I by the iodine monochloride method.

[b] The animals became azotemic in 48 hours with the blood urea nitrogen rising from 30 to 134 mg/100 ml, but they remained clinically well and active.

to the liver and other organs was not significantly different from that in normal animals. Since the excess polymer in the serum was not equal to the amount which the kidneys would have extracted had they been present, the remainder must have been evenly distributed among the other body compartments. These findings show either that the liver was already functioning at its maximum capacity for concentrating D-polymer or that some heterogeneity existed in the injected molecules such that the kidney rapidly extracted a certain population which was not readily taken up by the liver. There is some evidence for biological heterogeneity in synthetic polypeptides (GILL et al., 1964b), but it did not significantly affect serum elimination or tissue localization, since both were the same for the D-polypeptide when used before or after centrifugation at 105,000 × G for 120 minutes.

In summary, the marked differences in metabolic behavior between the isomeric polypeptides appear to be the major factor in the disparate immunogenic potencies of the polypeptides. The liver is probably the major site of catabolism of the polypeptides, and the kidney is the main storage depot for the D-polypeptide. Since the D-polymer can be degraded to some extent, there may be D-proteases and D-peptidases in the rabbit. There is some evidence for the existence of such enzymes in the finding of an enzyme capable of degrading poly(D-lysine) in the allantoic fluid of the chick embryo (TSUYUKI et al., 1956b) and in pancreatic extracts (TSUYUKI et al., 1956a).

Under some circumstances antigenic competition can occur between the L- and D-isomers of a polypeptide when they are administered in Freund's complete adjuvant. MAURER and PINCHUCK (1968) reported that immunization with 60 mg of a D-amino acid polymer in complete Freund's adjuvant depressed the response to the L-enantiomorph administered later in complete Freund's adjuvant. On the other hand, neonatal animals given a series of intraperitoneal and intravenous injections of D-amino acid polymers (130 mg in solution) showed a normal response to subsequent immunization with the L-enantiomorph in Freund's complete adjuvant.

Studies of the metabolic fate and organ degradation of poly[$\gamma$(D-glutamic acid)] from the capsule of the anthrax bacillis were undertaken by Goodman and his colleagues (ROELANTS et al., 1969a, 1969b) in an attempt to correlate immunogenicity with the metabolic fate of the antigen. The antigen was studied in three forms: soluble, alum-precipitated, and complexed with methylated bovine serum albumin (MeBSA). Only the complex with MeBSA in complete Freund's adjuvant elicited an antibody response in rabbits. The metabolic fate of poly[$\gamma$(D-glutamic acid)] following intravenous injection into rabbits is summarized in Table 5. All three forms of the antigen were rapidly cleared from the plasma, especially the soluble form. The soluble antigen and the MeBSA complex were degraded slowly and incompletely, whereas the alum-precipitated form was degraded rapidly and extensively. Only a homogenate of liver degraded poly[$\gamma$(D-glutamic acid)]. The soluble and alum-precipitated forms were retained mainly in the liver, and some of the alum-precipitated form was also retained in the glomeruli and periglomerular regions of the kidney. The MeBSA complex was retained mainly in the kidney where it was in the glomeruli and the proximal convoluted tubules. In the spleen, the alum-precipitated antigen and the MeBSA complex were retained in the red pulp where they were associated with large mononuclear cells, presumably macrophages. Both forms of the antigen showed scattered localization in the white pulp, and the MeBSA complex also showed focal localization in the white pulp.

The clearance from the plasma, the urinary excretion, and the total amount of polymer retained following injection of the three forms of the antigen into immunized rabbits were approximately the same as in normal animals. The major differences lay in the organ retention pattern: the MeBSA complex was retained mainly in the liver (50%), whereas only a small amount was present in the kidney (5%). In the immunized animals, the antigen retained in the spleen following the injection of soluble or alum-precipitated poly [$\gamma$(D-glutamic acid)] was extensively degraded, whereas the polymer in the animals immunized with the MeBSA complex was approximately 60% intact.

These studies showed that the metabolism of poly[$\gamma$(D-glutamic acid)] occurred mainly in the liver and that the main storage depots were the liver and the kidney. The relative proportions of polypeptide stored in these organs depended upon the form in which the antigen was presented and whether the animal had any circulating antibody. The only finding that suggested a correlation with immunogenicity was the retention of intact antigen in the spleen following immunization with the MeBSA complex, which was the only

Table 5. *The metabolic fate and organ*

| Form of poly [$\gamma$(D-glutamic acid)] used for immunization | Metabolic fate of tritiated poly [$\gamma$(D-glutamic acid)] intravenously into normal rabbits | | | | | |
|---|---|---|---|---|---|---|
| | antibody response[b] ($\mu$g Ab/ml) | plasma | | urinary excretion | | |
| | | 2 hrs. (%) | 3 wks. (%) | 2 days (%) | 1 wk. (%) | 3 wks. (%) |
| Soluble | 0 | 1 | <1 | 10 | 15 | 15 |
| Alum-precipitated | 0 | 7 | <1 | 25 | 65 | 85 |
| MeBSA-complexed | 30—80 | 7 | <1 | 5 | 15 | 15 |

[a] Roelants et al. (1969a, 1969b).

[b] Following intradermal or subcutaneous immunization with the antigen in complete Freund's adjuvant.

form in which the antigen elicited an antibody response. The retention of antigen in lymphoid tissue in sensitized animals has been pointed out by several groups, but its relationship to immunogenicity is, as yet, unclear (Cohen et al., 1966; Humphrey and Frank, 1967; Nossal et al., 1968a, 1968b; Miller et al., 1968; McConahey et al., 1968).

## Studies in Mice

The metabolic fate of linear, enantiomorphic polypeptides (Table 6) has been studied in mice by Janeway, Humphrey and Sela (Janeway and Sela, 1967; Janeway and Humphrey, 1968, 1969). Poly(DGlu$^{51}$DAla$^{40}$DTyr$^{9}$) (No. 247) and poly(Glu$^{49}$Ala$^{43}$Tyr$^{8}$) (No. 253) were labeled with $^{125}$I by the chloramine T method (Greenwood et al., 1963), and 5 $\mu$g were injected in saline into the hind footpads of (CBA$\times$C57)F1 hybrid mice of both sexes. The results of these experiments are summarized in Tables 6 and 7. The D-polymer was broken down 22 times more slowly than the L-polymer; in addition, some of the intact D-polymer was excreted in the urine. After footpad injection, 200 to 1,000 times more D-polymer was retained in the draining lymph nodes and spleen than was the case with the L-polymer. The major sites for retention of the D-polypeptide were the liver and kidney; in the latter organ the polymer localized in the proximal convoluted tubules. Radioautographs of the draining lymph nodes showed that the D-polymer was almost exclusively in macrophages and that the L-polymer was initially in macrophages and increasing also in germinal centers. Similar differences in the metabolism of the D- and L-polypeptides were seen in experiments using *degradation of poly [γ(D-glutamic acid)]*[a]

| injected | | | Degradation of poly [γ(D-glutamic acid)] by organ homogenates[d] | | | | State of polymer in the spleens of immunized rabbits |
|---|---|---|---|---|---|---|---|
| organ localization (3 weeks) | | | | | | | |
| liver (%) | kidney (%) | spleen (%) | liver | kidney | spleen or lymph node | serum | |
| 75 | 4 | 1 | ++++ | 0 | 0 | 0 | extensively degraded |
| 12 | 1 | 0.1 | | | | | extensively degraded |
| 11[c] | 54[c] | 17 | | | | | ca. 60% intact |

[c] In immunized rabbits, 50% of the retained polypeptide was in the liver and 5% was in the kidney. Other localization patterns with the MeBSA-complexed antigen were similar to those observed in normal rabbits.

[d] Incubated at 37° for 48 hours. Qualitative estimation ranges from 0 to ++++.

the same dose of antigen emulsified in complete Freund's adjuvant. Thus, the results of the metabolic studies in mice were quite similar to those in rabbits.

A similar metabolic study was carried out using 1 μg of poly(DGlu$^{51}$DAla$^{40}$DTyr$^{9}$) (No. 247) injected intraperitoneally in saline into newborn mice (JANEWAY and HUMPHREY, 1969). The patterns of organ localization were essentially the same as those in adults. In addition, the susceptibility of newborn mice to the induction of tolerance by the D-polypeptide was similar to the susceptibility of adults on a weight basis.

The localization of antigen in adult (CBA×C57)F1 hybrid mice of both sexes was studied in detail by McDEVITT et al. (1966) and by HUMPHREY et al. (1967). The antigens used in these studies were poly(Tyr$^{8}$Glu$^{15}$)-poly(DLAla$^{73}$)—poly(Lys$^{4}$) (No. 509) labeled with $^{125}$I or similar antigen preparations labeled with $^{125}$I and/or with tritium. When a single primary injection of 10 μg antigen in saline was given into the hind footpads, approximately 1% to 2% was retained at the injection site after 24 hours, and much smaller amounts were retained by the lymph nodes, liver, spleen, lungs and kidneys. Most of the $^{125}$I was excreted in the urine after the first 24 hours. Radioautographic study of the lymph nodes showed that there was marked retention of the antigen in the subcapsular sinus and in the medullary areas; the dense cortex and the intermediate zone showed very little label. Localization of the antigen over the germinal centers was very slight but definite: it was minimal at 12 to 24 hours, more marked at 3 days, and quite definite, but still light, at 7 and 14 days.

Table 6. *The physical chemical and biological properties of synthetic polypeptide antigens given for*

| Formula | Polypeptide antigens | | | |
|---|---|---|---|---|
| | No. | type | molecular weight | weight percent tyrosine |
| poly(DGlu$^{51}$DAla$^{40}$DTyr$^{9}$) | 247 | linear | 19,700 | 13.7 |
| poly(Glu$^{49}$Ala$^{43}$Tyr$^{8}$) | 253 | linear | 23,000 | 11.7 |
| poly(DGlu$^{60}$DLys$^{34}$DTyr$^{6}$) | 251 | linear | 44,000 | 7.2 |
| poly(Glu$^{60}$Lys$^{34}$Tyr$^{6}$) | 252 | linear | 61,000 | 7.2 |
| poly(DGlu$^{49}$DAla$^{43}$DTyr$^{8}$) | 236 | linear | 33,800 | 12.8 |
| poly(Tyr$^{8}$Glu$^{15}$)-poly(DLAla$^{73}$)—poly(Lys$^{4}$) | 509 | branched | 232,000 | 14.3 |
| poly(Tyr, Glu)-poly(DLAla)—poly(Lys) | | branched | 45,000 | 10.6 |
| poly(Tyr, Glu)-poly(DLAla)—poly(Lys) | 594 | branched | 20,000 | 4.6 |
| poly(DTyr$^{6}$DGlu$^{4}$)-poly(DPro$^{87}$)—poly(DLys$^{3}$) | 713 | branched | 171,000 | 9.0 |
| poly(DTyr$^{2}$DGlu$^{3}$)-poly(Pro$^{92}$)—poly(Lys$^{3}$) | 715 | branched | 225,000 | 2.9 |
| poly(Glu$^{58}$Lys$^{36}$Tyr$^{6}$) | 2 | linear | 70,000 | 7.5 |
| poly(DGlu$^{55}$DLys$^{39}$DTyr$^{6}$) | 1 | linear | 93,000 | 7.5 |

[a] Estimated and graded on a scale of 0 to ++++.

In a second experiment, the localization of antigen given as a primary injection in complete Freund's adjuvant was studied. A large amount of antigen was retained at the injection site, and the pattern of tissue localization was similar to that seen with the injection of the antigen in saline, but the localization occurred somewhat more slowly and was quantitatively less intense. One difference was the existence of marked aggregates of radio-labeled antigen within or on the walls of dilated lymphatics in the nodes; this finding was apparently due to the presence of antigen-containing oil droplets. There was no evidence for the specific retention of antigen in antibody-producing cells.

A third experiment explored the localization of antigen in the secondary response: the radioactive antigen was injected in saline following a primary dose of unlabeled antigen in Freund's complete adjuvant. The amounts of radio-labeled antigen retained were somewhat higher than the amounts retained at a corresponding time after the injection of the antigen in saline into unsensitized animals. There was a striking change in localization, however, with an intense concentration of antigen within all of the germinal centers of the lymphoid follicles. Again, there was no detectable antigen in plasma cells. Similar studies with $^{125}$I-labeled hemocyanin showed a pattern of distribution very similar to that seen with the synthetic polypeptide. Therefore, the major difference in antigen metabolism between primed and unprimed animals was *used for metabolic studies in the mouse. The data for the antigens used in rabbits are comparison*

| Qualitative estimate of biological properties[a] | | | | References |
|---|---|---|---|---|
| immunogenicity | induction of tolerance | persistence in the animal | | |
| | | serum | tissues | |
| + | ++++ | + | +++ | JANEWAY and HUMPHREY (1968) |
| +++ | 0 | 0 | 0 | JANEWAY and HUMPHREY (1968) |
| 0 to + | | + | +++ | JANEWAY (1969a); JANEWAY and SELA (1967) |
| | | 0 | 0 | JANEWAY (1969a) |
| + | +++ | + | +++ | MEDLIN et al. (1970b) |
| + | | | +++ | MCDEVITT et al. (1966) |
| | | | +++ | HUMPHREY et al. (1967) |
| 0 to + | | | +++ | HUMPHREY et al. (1967) |
| ++ | ++ | + | +++ | MEDLIN et al. (1970a) |
| ++ | +++ | ++ | ++ | MEDLIN et al. (1970a) |
| +++ | 0 | 0 | 0 | GILL et al. (1965)[b] |
| + | +++ | 0 | ++++ | GILL et al. (1965)[b] |

[b] In the rabbit.

the localization of antigen in the germinal centers of the lymphoid follicles in the primed animals. The significance of this observation for the induction or the maintenance of the immunological response is not clear, however.

MEDLIN, HUMPHREY and SELA (1970a, 1970b) studied the immunogenicity and metabolism of the branched polypeptides poly (DTyr$^6$DGlu$^4$)-poly (DPro$^{87}$)—poly (DLys$^3$) (No. 713) and poly (DTyr$^2$DGlu$^3$)-poly (Pro$^{92}$)—poly (Lys$^3$) (No. 715) (Table 6) in adult (CBA×C57) F1 hybrid mice of both sexes. Both polypeptides were more immunogenic than linear polymers of D-amino acids, and a relatively high primary response was seen following the injection of the polymers in saline or in Freund's complete adjuvant. The authors ascribed the greater immunogenicity of the branched polypeptides to their higher molecular weight and greater degree of complexity. There was clear evidence of an increased secondary response after small primary doses administered in adjuvant and in the case of poly (DTyr$^2$DGlu$^3$)-poly (Pro$^{92}$)—poly (Lys$^3$) (No. 715), following primary injection in saline also. Such an anamnestic response was seen with linear polymers only after a long interval between the first course of immunization and the booster injection (JANEWAY, 1969b).

The dose dependence of the antibody response for both antigens was different from that described for other D-amino acid polypeptides in the mouse and in the rabbit. Within the dose range used, the larger the quantity of the

Table 7. *The metabolic fate of enantiomorphic synthetic polypeptide antigens in (CBA × C57)F1 hybrid mice*[a]

| Location | Percentage of injected polypeptide retained in various locations | | | |
|---|---|---|---|---|
| | 3 days | 7 days | 14 days | 21 days |
| *poly(DGlu$^{51}$DAla$^{40}$DTyr$^{9}$) (No. 247)*[b, c] | | | | |
| Liver | 13 | 15 | 11 | 12 |
| Kidney | 32 | 17 | 9 | 12 |
| Spleen | 1 | 1 | 1 | 1 |
| Draining lymph node | 6 | 5 | 2 | 2 |
| Urine | 20 | 25 | 35 | 45 |
| *poly(Glu$^{49}$Ala$^{43}$Tyr$^{8}$) (No. 253)*[b] | | | | |
| Liver | 0.3 | 0.08 | 0.02 | 0.02 |
| Kidney | 0.1 | 0.04 | 0.01 | 0.007 |
| Spleen | 0.02 | 0.003 | 0 | 0 |
| Draining lymph node | 0.03 | 0.01 | 0.006 | 0.006 |
| Urine | 75 | 90 | 100 | 100 |

[a] JANEWAY and HUMPHREY (1968). The polypeptides were labeled with $^{125}$I by the chloramine T method.

[b] Polymer in saline (5 μg) was injected into the hind footpads. A similar distribution pattern was seen when the polymer was incorporated into complete Freund's adjuvant.

[c] Essentially the same results were obtained in newborn mice (JANEWAY and HUMPHREY, 1969).

branched polymers given, the greater the primary response. A striking feature of the response to poly(DTyr$^{6}$DGlu$^{4}$)-poly(DPro$^{87}$)—poly(DLys$^{3}$) (No. 713) administered intravenously over a dose range of 0.01 to 100 μg was the presence of a primary antibody response without the induction of immunological memory or of paralysis. On the other hand, poly(DTyr$^{2}$DGlu$^{3}$)-poly(Pro$^{92}$)—poly(Lys$^{3}$) (No. 715) given in similar doses demonstrated a primary response which could then be followed by a secondary response or by partial or complete paralysis, depending upon the dose of antigen given initially.

The degradation of poly(DTyr$^{2}$DGlu$^{3}$)-poly(Pro$^{92}$)—poly(Lys$^{3}$) (No. 715) in vivo was somewhat faster than that of poly(DTyr$^{6}$DGlu$^{4}$)-poly(DPro$^{87}$)—poly(DLys$^{3}$) (No. 713); the cumulative $^{125}$I excreted by 24 days was 50% and 25%, respectively. The initial phase of degradation was rapid, and nearly all the radioactivity excreted was soluble in trichloroacetic acid. After the first few days, however, 40% of the excreted $^{125}$I was bound to macromolecular material. The majority of the retained antigen was in the liver and the spleen, but relatively little was retained in the kidney, since the polymers were too big to be filtered through the glomerulus and absorbed by the tubule. A significant amount of polymer was excreted in the bowel where it was slowly degraded by feces; up to 15% of the retained antigen was associated with the intestinal tract. The polymers could also be degraded by urine, but not by serum. Simultaneous injection of either poly(DTyr$^{6}$DGlu$^{4}$)-poly(DPro$^{87}$)—poly(DLys$^{3}$) (No. 713) or poly (DTyr$^{2}$DGlu$^{3}$)-poly(Pro$^{92}$)—poly(Lys$^{3}$) (No. 715) and poly(DGlu$^{49}$DAla$^{43}$DTyr$^{8}$) (No. 236) showed that administration of the linear polymer did not affect the clearance and degradation of the branched polymer.

The organ localization of both branched polypeptides was essentially the same. In the spleen, they were in the phagocytes lining the sinusoids of the red pulp. Localization in the white pulp was restricted to macrophages, and there was little or no evidence for the presence of polypeptide in the germinal centers of lymphoid follicles. In lymph nodes, the polymers were sequestered in the macrophages of the medulla and the subcapsular sinus. The polypeptide in the liver was in the Kupffer cells lining the sinusoids. The kidney showed a small amount of polymer in the glomeruli and a very small amount in the proximal convoluted tubules. In the gastrointestinal tract, the retained polypeptide was associated with macrophages in the wall of the bowel. One excretory mechanism for the larger polymers, which could not be effectively excreted through the kidney, was phagocytosis and removal of the polymer-containing phagocytes through the bowel wall into the feces. This mechanism could not account for a major portion of the polymer excretion, however, and it was inferred that there must be extensive polymer degradation in the tissues. The amounts of the two branched polymers retained in the tissues, including the macrophages, were comparable (25% to 50% of the injected dose), and it was approximately the same as the amount of poly(DGlu$^{51}$DAla$^{40}$DTyr$^{9}$) (No. 247) retained in the tissues (20 to 60%). The major exception was the persistence of more poly(DTyr$^{2}$DGlu$^{3}$)-poly(Pro$^{92}$)—poly(Lys$^{3}$) (No. 715) in the circulation, and this was postulated to be the basis of the greater ability of this polypeptide to induce tolerance.

In summary, after intravenous administration of poly(DTyr$^{6}$DGlu$^{4}$)-poly(DPro$^{87}$)—poly(DLys$^{3}$) (No. 713), poly(DTyr$^{2}$DGlu$^{3}$)-poly(Pro$^{92}$)—poly(Lys$^{3}$) (No. 715) and poly(DGlu$^{49}$DAla$^{43}$DTyr$^{8}$) (No. 236), the polypeptides were removed rapidly from the blood and taken up by macrophages. The poly(DTyr$^{6}$DGlu$^{4}$)-poly(DPro$^{87}$)—poly(DLys$^{3}$) (No. 713) was retained intracellularly so well that it vanished from the circulation, whereas poly(DTyr$^{2}$ DGlu$^{3}$)-poly(Pro$^{92}$)—poly(Lys$^{3}$) (No. 715) concentrations in the blood fell somewhat more slowly and remained present for a long period of time. Poly(DGlu$^{49}$DAla$^{43}$DTyr$^{8}$) (No. 236) also remained present at low but detectable levels for several weeks. Thus, the capacity of the polymers to induce tolerance appeared to correlate with the level and duration of their persistence in the tissues and, especially, in the circulation (Table 6). The branched polymers injected in saline elicited readily detectable circulating antibody, whereas the linear polymer poly(DGlu$^{49}$DAla$^{43}$DTyr$^{8}$) (No. 236) did not; the reason for this finding remains unexplained. Finally, these experiments showed that retention of antigens within macrophages did not, by itself, increase their immunogenicity.

## Studies in Guinea Pigs

The metabolism of various conjugates of poly(L-lysine) and poly(D-lysine) in guinea pigs was investigated by LEVINE and BENACERRAF (1964). Studies in vitro utilized splenic preparations from responder and non-responding guinea pigs, which were made from frozen spleens extracted with phosphate-buffered saline and stored frozen until used. The enzymatic degradation of the polypeptide conjugates was performed in the presence of cysteine at pH 4.9 and at 37° for 20 hours; the concentration of the various conjugates in the digestion

Table 8. *The immunogenicity of various conjugates of poly(Lys) and poly(DLys) in guinea pigs and their degradation in vitro and in vivo*[a]

| Polymer | Immunogenicity in strain 2 and responder random bred guinea pigs | In vitro degradation by splenic enzymes | | In vivo degradation and urinary excretion (after 3 days) | |
|---|---|---|---|---|---|
| | | responder[b] | non-responder[b] | responder | non-responder |
| F-poly(Lys) (h) | + | degraded | degraded | | |
| F-poly(LysS) (h) | ± | degraded | degraded | | |
| DNP-poly(Lys) (l) | + | | | 70—90% | 60—80% |
| DNP-poly(LysS) (l) | — | | | | 80—90% |
| F-poly(DLys) (l) | — | none | none | | |
| DNP-poly(DLys) (l) | — | | | | <1% |

[a] LEVINE and BENACERRAF (1964). The symbols are: (h), relatively high molecular weight (approximately 80,000); (l), relatively low molecular weight (20,000—40,000); LysS, succinylated lysine; F, fluorescein; and DNP, 2,4-dinitrophenyl.

[b] The chromatographic patterns of the degradation products were the same for a given antigen in the responders and non-responders.

Table 9. *Response to various hapten-polycation complexes in guinea*

| Guinea pig strain | Response defined by immunization with BPO-poly (Lys) | DNP-poly (Lys) | | |
|---|---|---|---|---|
| | | μg Ab/ml | DH | μg Ab/ml (BSA complex) |
| Strain 2 | responder | 820 | +++ | |
| Random bred | responder | 1,600 | ++++ | |
| Random bred | non-responder | 0 | 0 | 1,180 |
| Rank order of immunogenicity in responders | | | 1 | |
| Relative rate of trypsin hydrolysis | | | 2 | |

[a] LEVINE (1969). All polymers had a relatively high molecular weight.

mixture was 2 mg/ml. The digestion products were then analyzed by paper chromatography or by starch gel electrophoresis (Table 8). Fluorescein-labeled polylysine or succinylated polylysine was degraded by extracts from both responder and non-responder spleens, and the peptide patterns were approximately the same. The succinylated conjugate was less immunogenic than the parent compound, but this was not reflected in the degradation patterns. The in vivo studies employed intraperitoneal injections of tritiated DNP-polylysine conjugates and detection of the radioactive label in the urine over a period of three days. From 70% to 90% of the DNP-polylysine was degraded by the responder guinea pigs, and 60% to 80% was degraded by the non-responders. Succinylation of DNP-polylysine did not significantly alter its degradation by non-responder guinea pigs, although the DNP-polylysine alone was immunogenic. Finally, poly (D-lysine) was not degraded in vitro by splenic extracts nor was it degraded in vivo. Therefore, there appeared to be no direct correlation between the in vitro or in vivo degradation of the various polylysine antigens and their ability to elicit an immune response. However, in the cases where the polylysine derivatives were immunogenic, they could all be degraded, whereas the non-immunogenic poly(D-lysine) could not be degraded. So, degradation may play a role in immunogenicity, but the relationship is not a simple and direct one.

The dinitrophenyl derivatives of various polycations showed markedly different immunogenic potencies (LEVINE, 1969). The antibody response to DNP-derivatives of polylysine, polyarginine, polyhomoarginine and polyornithine are shown in Table 9. Strain 2 guinea pigs responded with a progressively decreasing amount of antibody to these derivatives, and the responders among random bred guinea pigs showed the same rank order of response. The non-responders of the random bred strain responded to the various conjugates when they were immunized with the polycation complexed with bovine serum albumin (BSA). However, this procedure destroyed the rank order of the im-

*pigs and the susceptibility of these compounds to trypsin hydrolysis*[a]

| DNP-poly(Arg) | | | DNP-poly(Harg) | | | DNP-poly(Orn) | | |
|---|---|---|---|---|---|---|---|---|
| μg Ab/ml | DH | μg Ab/ml (BSA complex) | μg Ab/ml | DH | μg Ab/ml (BSA complex) | μg Ab/ml | DH | μg Ab/ml (BSA complex) |
| 440 | +++ | | 130 | ++ | | 0 | 0 | |
| 400 | +++ | | 120 | ++ | | 20 | 0 | |
| 0 | 0 | 470 | 0 | 0 | 480 | 0 | 0 | 500 |
| | 2 | | | 3 | | | 4 | |
| | 1 | | | 3 | | | 4 | |

The symbols are: BPO, benzoylpenicillin; DH, delayed hypersensitivity and Harg, homoarginine. DH graded on a scale of 0 to ++++.

mune response, and the guinea pigs responded to the polylysine conjugate much better than to the other polycation conjugates, which all elicited approximately the same amount of antibody.

In an effort to determine whether the rank order of immunogenicity was a function of the ability of the polymers to be hydrolyzed, the susceptibility of the various conjugates to hydrolysis by trypsin was tested. The rank order of hydrolysis was similar, but not the same: DNP-polyarginine was hydrolyzed faster than DNP-polylysine. Thus, there was no clear correlation between the susceptibility to trypsin hydrolysis and the rank order of immunogenicity.

## Radioactive Labels for Metabolic Studies

The establishment of the validity of the iodine label is important, since it is used so frequently in metabolic studies. On the basis of the observation that slices of liver and kidney are capable of deiodinating thyroxin (ALBRIGHT et al., 1954) and di-iodo-tyrosine (TONG et al., 1954) in vitro, deiodination of a labeled antigen without affecting the macromolecule itself has been postulated to occur in vivo. There is no evidence that this does occur, however, and metabolic studies in rabbits (CARPENTER et al., 1967) showed that the very active deiodinases of the kidney did not affect retained, iodine-labeled D-polymer. The experiments of LAWS (1952) on the metabolism of $^{131}$I-labeled albumin demonstrated the presence of small amounts of labeled peptides in addition to free iodide in the urine. This finding provides additional evidence that the protein was degraded and not simply deiodinated. Since animals provided with $^{131}$I alone did not excrete labeled peptides (CARPENTER et al., 1967), it seems unlikely that any significant reutilization of iodine occurred in animals whose thyroid function had been adequately blocked. Others have failed to show iodine reutilization as well (COHEN et al., 1956; WALTER et al., 1957).

Experiments by JANEWAY (1969a) on the validity of the iodine label utilized the hapten 4-hydroxy-3-iodo-5-nitrophenylacetic acid (NIP) containing $^{131}$I conjugated to poly(DGlu$^{60}$DLys$^{34}$DTyr$^{6}$) (No. 251) or to poly (Glu$^{60}$Lys$^{34}$Tyr$^{6}$) (No. 252), which were then labeled with $^{125}$I. Mice were injected intraperitoneally with a solution of either NIP-polypeptide conjugate, and the catabolism of the polypeptide was followed. The catabolism was essentially the same when measured either by the excretion of $^{125}$I or of $^{131}$I (Fig. 8). Thus, there was no evidence for a selective hydrolysis of the hapten from the polypeptide carrier, and both markers were valid labels.

The validity of the external $^{125}$I label and the internal tritium label was tested by using poly(Tyr, Glu)-poly(DLAla)—poly(Lys) labeled with $^{125}$I by the chloramine T method and poly (Tyr, Glu)-poly (DLAla)—poly (Lys) (No. 594) synthesized with tritiated alanine and by using a doubly-labeled polymer, the tritiated polypeptide labeled with $^{125}$I (HUMPHREY et al., 1967). In (CBA × C57) F1 hybrid mice of both sexes, primary and secondary responses to the polypeptides were studied. The tritiated polymer was less immunogenic than the iodinated polypeptide, and this was attributed to its lower tyrosine content and smaller molecular weight (Table 6). The localizations of the tritiated, iodinated and iodinated-tritiated polypeptides were quite similar in the primary and secondary responses, and they were essentially the same as those described by McDevitt et al. (1966). Again, no antigen was found in antibody-producing cells. Although the antigen localization patterns were the same with both labels, the ratio of tritium to $^{125}$I in the retained, doubly-labeled antigen varied over a two to five-fold range, indicating a selective loss of iodine. This finding was ascribed to the fact that the terminal portions of the branched polymer, which contained the tyrosine, were relatively short compared with

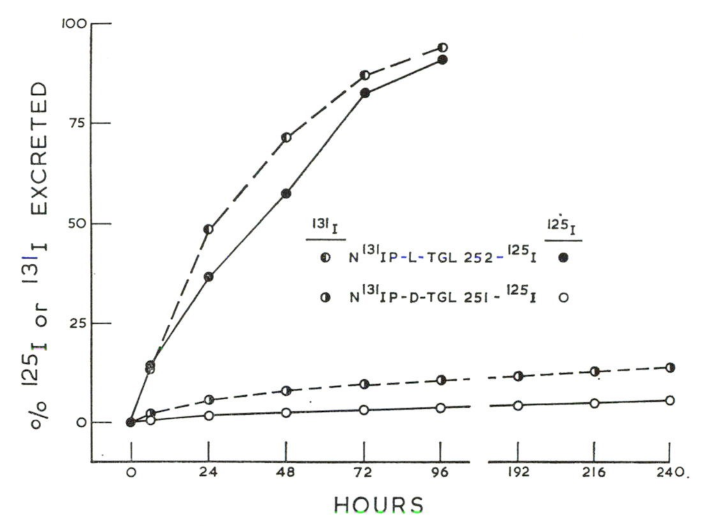


Fig. 8. The cumulative excretion of $^{131}$I and $^{125}$I by mice injected with 95 μg of poly (Glu$^{60}$Lys$^{34}$Tyr$^{6}$) (No. 252) or with 88 μg of poly(DGlu$^{60}$DLys$^{34}$DTyr$^{6}$) (No. 251). Both polypeptides were labeled with $^{125}$I and coupled with the NIP hapten containing $^{131}$I (N$^{131}$IP). (Janeway, 1969a)

the inner poly(DLAla) sequences and were susceptible to L-amino acid peptidases present in cells and tissue fluids. Such enzymes would not split the bond between the adjacent alanine residues when one of them had the D-configuration. Hence, the portion of the molecule containing the tritium would be indigestable, and the tritium label would be selectively retained in the tissues for a long period of time. Direct confirmation of this hypothesis was obtained by showing that the tyrosine-containing sidechains of the polypeptide could be cleaved when a 4% solution of polypeptide was incubated for 2 to 15 hours at 35° with normal mouse or rabbit serum. Thus, the authors concluded that radio-iodine can be regarded without qualification as a valid marker for antigens.

There was no indication for the reutilization of tritium from the poly (DL-alanine) and, in addition, mice injected with tritiated DL-alanine did not show any specific organ localization of the amino acid (Humphrey et al., 1967). These findings suggest that the tritium label was not reutilized. However, since the tritium was present in a form which was not very susceptible to degradation, the question of the reutilization of tritium incorporated into a polypeptide and its influence on the validity of internal labeling with tritium as a radioactive marker has not been resolved.

The use of tritium-labeled poly[$\gamma$(D-glutamic acid)] apparently did not cause any significant problems with reutilization of the tritium (ROELANTS et al., 1969a, 1969b). The evidence against utilization of labile tritium atoms or of degradation products of the polypeptide was inferred from the following observations: (a) electrophoretic identity of all the radioactivity excreted in the urine or retained in the spleen with the electrophoretic patterns of glutamic acid; (b) the solubility in trichloroacetic acid of high molecular weight radioactive material isolated from the spleen — a property peculiar to high molecular weight $\gamma$(D-glutamic acid) polypeptides; and (c) the resistence of the high molecular weight radioactive material to proteolytic digestion under conditions which extensively degraded other proteins, including poly[$\gamma$(L-glutamic acid)]. These findings provide further evidence that the tritium label may be a useful radioactive label for metabolic studies. Nonetheless, they utilized an unusual antigen, and caution must be exercised in extrapolating to L-amino acid polypeptides and proteins.

The use of ferrocene as a label resulted in serum elimination and urine excretion curves which were the same as those for the iodinated polymers (CARPENTER et al., 1967). However, the organ localization of the polymers was influenced by the ferrocene label, and a significant amount of reutilization of iron occurred with incorporation into hemoglobin and into unknown materials in the liver and the kidneys. Ferrocene is also an excellent hematinic (MADINAVEITIA, 1965), and its incorporation into liver has been previously documented (DRATZ et al., 1964). For these reasons it is not as satisfactory for tissue studies as iodine, which appears to have the unique advantage of not being reincorporated into other compounds.

The $^{35}$S-sulfanilate, which GARVEY and CAMPBELL (1956, 1957) have used extensively, is similar to iodine in that it was not significantly reutilized but remained bound to the injected antigen, even after the latter was degraded to small peptides. HAUROWITZ and WALTER (1955) using the $^{35}$S-azophenylsulfonate label also found that retained $^{35}$S-radioactivity persisted essentially unchanged and was bound to a protein fraction. Results with the $^{14}$C-azobenzoate label (FRIEDBERG et al., 1955) were similar, and in both cases double-label experiments showed that $^{35}$S and $^{14}$C persisted in tissues much longer than $^{131}$I. However, it is not known to what extent the retained azophenylsulfonate and azobenzoate labels represent the persistence of antigen. Other studies (FLEISCHER et al., 1959; WALTER et al., 1961) show that internal labels such as $^{35}$S- or $^{14}$C-amino acids lead to a considerable amount of reutilization, and incorporation into tissue protein may occur.

In conclusion, external labels are probably the most reliable of the available isotopic labels for macromolecules, and of these, iodine provides the best tracer for metabolic studies.

## Discussion

When an animal is injected with an antigen, there is a balance between immunological stimulation and paralysis which depends upon the chemistry of the antigen and the genetic background of the host. This balance varies for each antigen, and the chemical properties of the antigen set the level and range of dosage that can be used in stimulating an antibody response. The antigen acts intact, and the role of antigen catabolism is to function in concert with the original dose to regulate the amount of antigen available to stimulate antibody formation. A poor immunogen is probably a molecule that induces tolerance easily, but this proposition is a difficult one to test, since the mechanism of tolerance is not well understood.

One hypothesis to explain the action of antigen at the cellular level proposes that stimulation or tolerance depends upon the same sequence of events, which is governed by the mass action law and whose outcome depends upon multiple, interrelated equilibria. The amount of antigen that is available to stimulate immunocompetent cells depends upon the dose, the time over which it is given, and the rate at which the antigen is degraded in vivo. The ability of this effective concentration of antigen to stimulate an antibody response or to induce tolerance depends upon the genetically determined number of immunocompetent cells capable of reacting with the antigen and the binding affinity of the cellular receptors. The effector substances (antibodies or sensitized cells) which are induced following stimulation probably have a direct effect on the immunocompetent cell, or cells, in such a manner as to exert a negative feedback control. The population of immunocompetent cells capable of reacting with a given antigen and producing antibody could be increased by periodic stimulation with the appropriate amount of antigen; in this way tolerance could be avoided. In like manner, immunological memory may be due to the continued stimulation of the immunocompetent cell population by small amounts of antigen retained in the host. The retained antigen does not have to reside in the lymphoid tissue, but it may be released slowly into the circulation from a variety of tissues.

*Acknowledgment*

I wish to thank Professor Michael Sela for making several manuscripts available prior to publication; Dr. Charles A. Janeway, Jr. for allowing me to quote unpublished material and to publish Fig. 8 from his thesis; and the Williams and Wilkins Co., Baltimore, Maryland for allowing me to use Figs. 1—7 from the Journal of Immunology, volume 95 (Figs. 3 and 4) and volume 97 (Figs. 1—4, 6 and 7).

## References

Albright, E. C., Larson, F. C., Tust, R. H.: In vitro conversion of thyroxin to triiodothyronine by kidney slices. Proc. Soc. exp. Biol. (N.Y.) **86**, 137—140 (1954).

Campbell, D. H., Garvey, J. S.: Nature of retained antigen and its role in immune mechanisms. In: Advances in immunology, vol. 3, p. 261—313. New York: Academic Press 1963.

— — Localization and fate of foreign antigens in tissues. In: Immunological diseases, p. 18—31. Boston: Little, Brown & Co. 1965.

CARPENTER, C. B., GILL, T. J., III, MANN, L. T., JR.: Synthetic polypeptide metabolism. III. Degradation and organ localization of isomeric synthetic polypeptide antigens. J. Immunol. **98**, 236—250 (1967).
COHEN, S., HOLLOWAY, R. C., MATTHEWS, C., MCFARLANE, A. S.: Distribution and elimination of $^{131}$I- and $^{14}$C-labelled plasma proteins in the rabbit. Biochem. J. **62**, 143—154 (1956).
— VASSALLI, P., BENACERRAF, B., MCCLUSKEY, R. T.: The distribution of antigenic and non-antigenic compounds within draining lymph nodes. Lab. Invest. **15**, 1143—1155 (1966).
DRATZ, A. F., JR., COBERLY, J. C., GOLDSTEIN, J. H.: An exploratory study of $Fe^{59}$-labeled ferrocene as a carrier in tracer techniques. J. nucl. Med. **5**, 40—47 (1964).
FLEISCHER, S., LIETZE, A., WALTER, H., HAUROWITZ, F.: Conversion of serum proteins into tissue proteins. Proc. Soc. exp. Biol. (N.Y.) **101**, 860—863 (1959).
FRIEDBERG, W., WALTER, H., HAUROWITZ, F.: The fate in rats of internally and externally labelled heterologous proteins. J. Immunol. **75**, 315—320 (1955).
GARVEY, J. S., CAMPBELL, D. H.: Studies of the retention and properties of $S^{35}$ labeled antigen in livers of immunized rabbits. J. Immunol. **76**, 36—45 (1956).
— — The retention of $S^{35}$-labelled bovine serum albumin in normal and immunized rabbit liver tissue. J. exp. Med. **105**, 361—372 (1957).
GILL, T. J., III, DAMMIN, G. J.: Studies on synthetic polypeptide antigens. IV. The metabolic fate of the antigens. Biochim. biophys. Acta (Amst.) **56**, 344—348 (1962).
— GOULD, H. J., DOTY, P.: Role of optical isomers in determining the antigenicity of synthetic polypeptides. Nature (Lond.) **197**, 746—747 (1963).
— KUNZ, H. W., GOULD, H. J., DOTY, P.: Studies on synthetic polypeptide antigens. XI. The antigenicity of optically isomeric synthetic polypeptides. J. biol. Chem. **239**, 1107—1113 (1964a).
— — PAPERMASTER, D. S.: Studies on synthetic polypeptide antigens. XVIII. The role of composition, charge and optical isomerism in the immunogenicity of synthetic polypeptides. J. biol. Chem. **242**, 3308—3318 (1967).
— PAPERMASTER, D. S., MOWBRAY, J. F.: Metabolism of isomeric synthetic polypeptides. Nature (Lond.) **203**, 644—645 (1964b).
— — — Synthetic polypeptide metabolism. I. The metabolic fate of enantiomorphic polymers. J. Immunol. **95**, 794—803 (1965).
GREENWOOD, F. C., HUNTER, W. M., GLOVER, J. S.: The preparation of $^{131}$I-labelled human growth hormone of high specific radioactivity. Biochem. J. **89**, 114—123 (1963).
HAUROWITZ, F., WALTER, H.: Stability of an azoprotein hapten in the organism. Proc. Soc. exp. Biol. (N.Y.) **88**, 67—69 (1955).
HUGHES, W. L.: The major role of the kidney in catabolism of serum albumin. In: Proceedings of the 8th Annual Conference on the Nephrotic Syndrome, p. 22—30, ed. J. METCOFF (1956).
HUMPHREY, J. H., ASKONAS, B. A., AUZINS, I., SCHECHTER, I., SELA, M.: The localization of antigen in lymph nodes and its relation to specific antibody-producing cells. II. Comparison of iodine-125 and tritium labels. Immunology **13**, 71—86 (1967).
— FRANK, M. M.: The localization of non-microbial antigens in the draining lymph nodes of tolerant, normal and primed rabbits. Immunology **13**, 87—100 (1967).
JANEWAY, C. A., JR.: Studies on synthetic polypeptide antigens composed exclusively of D- or L-amino acids. Thesis submitted to the Harvard Medical School (1969a).
— Synthetic antigens composed exclusively of L- or D-amino acids. III. Anamnestic response to a synthetic polypeptide composed exclusively of D-amino acids. Immunology **17**, 715—721 (1969b).

JANEWAY, C. A., JR.: HUMPHREY, J. H.: Synthetic antigens composed exclusively of L- or D-amino acids. II. Effect of optical configuration on the metabolism and fate of synthetic polypeptide antigens in mice. Immunology **14**, 225—234 (1968).

— — The fate of a D-amino acid polypeptide [p(D-Tyr, D-Glu, D-Ala), 247] in newborn and adult mice. Relationship to the induction of tolerance. Israel J. med. Sci. **5**, 185—195 (1969).

— SELA, M.: Synthetic antigens composed exclusively of L- or D-amino acids. I. Effect of optical configuration on the immunogenicity of synthetic polypeptides in mice. Immunology **13**, 29—38 (1967).

LAWS, J. O.: The degradation of proteins labelled with radioactive iodine in control and sensitized rabbits. Brit. J. exp. Path. **33**, 354—358 (1952).

LEVINE, B. B.: Studies on the polylysine immune responder gene. The rank order of immunogenicity of dinitrophenyl conjugates of basic homopolyamino acids in guinea pigs. J. Immunol. **103**, 931—936 (1969).

— BENACERRAF, B.: Studies on antigenicity. The relationship between in vivo and in vitro enzymatic degradability of hapten-polylysine conjugates and their antigenicities in guinea pigs. J. exp. Med. **120**, 955—965 (1964).

MADINAVEITIA, J. L.: Ferrocenes as haematinics. Brit. J. Pharmacol. **24**, 352—359 (1965).

MAURER, P. H., PINCHUCK, P.: Effect of prior exposure to synthetic polymers of α-D-amino acids on the subsequent response to the enantiomorphic polymers of α-L-amino acids. Proc. Soc. exp. Biol. (N.Y.) **128**, 1112—1116 (1968).

MCCONAHEY, P. J., CEROTTINI, J.-C., DIXON, F. J.: An approach to the quantitation of immunogenic antigen. J. exp. Med. **127**, 1003—1011 (1968).

MCDEVITT, H. O., ASKONAS, B. A., HUMPHREY, J. H., SCHECHTER, I., SELA, M.: The localization of antigen in relation to specific antibody-producing cells. I. Use of a synthetic polypeptide [(T,G)-A--L] labelled with iodine-125. Immunology **11**, 337—351 (1966).

MCFARLANE, A. S.: Efficient trace-labelling of proteins with iodine. Nature (Lond.) **182**, 53 (1958).

MEDLIN, J., HUMPHREY, J. H., SELA, M.: Studies on synthetic polypeptide antigens derived from multichain polyproline. I. Immunogenic and tolerogenic capacity of antigens composed wholly or partly of D-amino acids. Folia Biol., **16**, 145 (1970a).

— — — Studies on synthetic polypeptide antigens derived from multichain polyproline. II. Metabolism and localization. Folia Biol., **16**, 156 (1970b).

MILLER, J. J., III, JOHNSEN, D. O., ADA, G. L.: Differences in localization of Salmonella flagella in lymph node follicles of germfree and conventional rats. Nature (Lond.) **217**, 1059—1061 (1968).

NOSSAL, G. J. V., ABBOT, A., MITCHELL, J.: Antigens in immunity. XIV. Electron microscopy autoradiography studies of antigen capture in the lymph node medulla. J. exp. Med. **127**, 263—275 (1968a).

— — — Antigens in immunity. XV. Ultrastructural features of antigenic capture in primary and secondary lymphoid follicles. J. exp. Med. **127**, 277—289 (1968b).

OLIVER, J., MACDOWELL, M.: Cellular mechanisms of protein metabolism in the nephron. VII. The characteristics and significance of the protein absorption droplets (hyaline droplets) in epidemic hemorrhagic fever and other renal diseases. J. exp. Med. **107**, 731—754 (1958).

PAPERMASTER, D. S., GILL, T. J., III, ANDERSON, W. F.: Synthetic polypeptide metabolism. II. The binding of synthetic polypeptides to serum proteins. J. Immunol. **95**, 804—809 (1965).

RICHTER, M., ZIMMERMAN, S., HAUROWITZ, F.: Relationship of antibody titer to presistence of antigen. J. Immunol. **94**, 938—941 (1965).

ROELANTS, G. E., SENYK, G., GOODMAN, J. W.: Immunochemical studies on the poly-γ-D-glutamyl capsule of bacillus anthracis. V. The in vivo fate and distribution in rabbits of the polypeptide in immunogenic and nonimmunogenic forms. Israel J. med. Sci. **5**, 196—208 (1969a).
— WHITTEN, L. F., HOBSON, A., GOODMAN, J. W.: Immunochemical studies on the poly-γ-D-glutamyl capsule of bacillus anthracis. VI. The in vivo fate and distribution in immunized rabbits of the polypeptide in immunogenic and nonimmunogenic forms. J. Immunol. **103**, 937—943 (1969b).
SHUSTER, S., JONES, J. H., FLYNN, G.: Renal tubular secretion of human plasma proteins and Bence-Jones protein. Brit. J. exp. Path. **44**, 145—150 (1963).
SOLOMON, A., WALDMANN, T. A., FAHEY, J. L., MCFARLANE, A. S.: Metabolism of Bence-Jones proteins. J. clin. Invest. **43**, 103—117 (1964).
TALMAGE, D. W., BAKER, H. R., AKESON, W.: The separation and analysis of labelled antibodies. J. infect. Dis. **94**, 199—212 (1954).
TONG, W., TAUROG, A., CHAIKOFF, I. L.: The metabolism of $I^{131}$-labeled diiodotyrosine. J. biol. Chem. **207**, 59—76 (1954).
TSUYUKI, E., TSUYUKI, H., STAHMANN, M. A.: Inhibition of mumps and influenza B virus multiplication by synthetic poly-D-lysine. Proc. Soc. exp. Biol. (N. Y.) **91**, 318—320 (1956a).
— — — The synthesis and enzymatic hydrolysis of poly-D-lysine. J. biol. Chem. **222**, 479—485 (1956b).
WALTER, H., FLEISCHER, S., HAUROWITZ, F.: Effect of different external labels on the metabolism of doubly labeled heterologous proteins. Arch. Biochem. **95**, 290—295 (1961).
— HAUROWITZ, F., FLEISCHER, S., LIETZE, A., CHENG, H. F., TURNER, J. E., and FRIEDBERG, W.: The metabolic fate of injected homologous serum proteins in rabbits. J. biol. Chem. **224**, 107—119 (1957).

Department of Chemical Immunology · The Weizmann Institute of Science
Rehovot, Israel

# Antibodies to Enzymes - A Tool in the Study of Antigenic Specificity Determinants

RUTH ARNON

With 7 Figures

Contents



## 1. Introduction

### a) Scope

The study of antigenic specificity determinants has long been a subject of interest. Since the early studies of LANDSTEINER (1945) on the specificity of serological reactions and through numerous recent investigations with synthetic antigens (SELA, 1966) as well as protein conjugates (BOYD, 1962; PRESSMAN and GROSSBERG, 1968), efforts have been aimed at the elucidation of the structural features which are characteristic of the sites against which the antibodies are formed. Antibodies can be produced against the majority of naturally occurring proteins as well as against many synthetic and artificial antigens. Whereas in the latter case the formation of antibodies against known structural groupings is elicited by defined specificity determinants, planted on carriers, in the case of naturally occurring proteins multiple different antigenic determinants are present on the same antigen molecule, and no information is available on the particular groups, or arrangements of groups, which elicit antibody production, or serve as points of recognition by the antibodies. Attempts have been made to identify such determinants by comparing the immunological cross-reaction between antigens of related structure, by investigation of the effects of chemical modification on the serological specificity, and by characterization of immunologically reactive fragments of the antigen (e.g. Crumpton, 1967; Atassi and Saplin, 1968; Shinka et al., 1962, 1967; Benjamini et al., 1964). It should be borne in mind, however, that the results of each such investigation depend not only on the antigen used but on the antiserum as well. The presence of many specificity determinants on the same antigen brings about inevitable heterogeneity in the antibodies, and different individual antisera will, therefore, differ in their potential capacity to react with the various antigenic sites of the homologous antigen. To overcome or bypass this difficulty, it is desirable to have means of differentiation between antisera which vary in the distribution of antibodies with distinct specificities towards different antigenic determinants on the same multideterminant antigen.

Enzyme-anti-enzyme systems offer the required flexibility for this type of investigation, because these antigens possess biological activity which resides in a limited area of the molecule. Antibodies specific towards this or related regions, inhibit the catalytic activity. Consequently, the contribution of different determinants to the immunological reactivity of an enzyme can be evaluated in the light of their relationship to its catalytic site. While this point will be discussed in detail in the next chapter, it may be illustrated here by one example: the totality of antibodies against papain has been found to be separable into two fractions, one of which contains highly efficient inhibitors of enzymic activity, whereas the other one lacks completely the capacity to inactivate the enzyme (Arnon and Shapira, 1967). These two fractions are undoubtedly reactive with distinct regions on the surface of the papain molecule, regions which are differently related to the catalytic site. Enzyme-anti-enzyme systems can thus serve as a tool in the study and localization of antigenic specificity determinants.

### b) Topics Approached by Immunoenzymological Studies

Concomitantly with its direct contribution to the study of antigenic determinants, the immunological approach to the study of enzymes has also assisted in the pursuit of several other problems arising in contemporary biology and enzyme research, and which derive from the antigenic multivalency of enzymes.

The question of biochemical evolution can serve as an excellent illustration for the fruitfulness of the immunoenzymological approach. Antibodies to specific enzymes can be employed, for example, in the search for enzymes of biological pathways which disappeared in the course of evolution, or to detect the extent of similarity between enzymes that persisted through the ages. In cases where the primary sequence of the enzymes is known, the elucidation of the immunological behaviour is feasible in precise molecular terms and should provide a sensitive probe of the surface conformation, in addition to the contribution in identifying antigenic determinants. The extensive studies on cytochrome *c* (e.g. MARGOLIASH et al., 1967) have indeed demonstrated the power of this analytical approach. They demonstrated that antisera prepared against any one of several cytochromes *c* (the complete amino acid sequence of which is known) cross-react to a varying extent with the proteins from over 25 other species. The study of such a cross-reaction between molecules that have minimal differences in their amino acid sequence has made it possible to localize and identify some of the antigenic determinants, and to prove that they involve certain amino acid residues and not others.

In many cases the evolutionary changes that occur in the molecular composition and structure of an enzyme may be very extensive although the catalytic activity and specificity are barely affected. By studying the immunological interactions of isofunctional enzymes isolated from different species, one might therefore expect cross-inhibition by the antibodies to accompany the cross-reaction, and this would imply greater similarity between the structures of the catalytic centers or related regions in the enzyme molecules than between other regions. In this case again conclusions about the localization of antigenic determinants could be drawn.

Structure-function correlation is one of the most important aspects in enzyme research, its goal being the elucidation of the structure of the catalytic site. Modification of enzymes has been a valuable means of determining which amino acids take part in the interaction with the substrate. Immunological studies on such modified enzymes can serve to define the nature of a particular modification; thus the correlation between enzymic and immunological activity may point to the role which the active site plays in the antigenic make-up of the enzyme molecule. In the case of ribonuclease, for example, chemical modification has been used to pinpoint three different amino acid residues which are essential for the maintenance of catalytic activity but play no part in the antigenic properties of the molecule (STARK, STEIN and MOORE, 1961; HIRS et al., 1961; BROWN, 1963). Similar studies performed with trypsin also indicated that modifications which affect the catalytic activity do not coincide with those changing the immunological reactivity (ARNON and NEURATH, 1970). On the other hand, numerous other cases have been reported where chemical modifications of enzymes simultaneously influenced both the enzymic and the immunological properties of the enzymes (e.g. HABEEB and ATASSI, 1969).

Many enzymes, chiefly animal proteases, are present within the tissues in the form of inactive precursors called zymogens or proenzymes. In this way they are prevented from exerting their destructive power on the protein components of the tissue in which they originate. Limited hydrolysis of some peptide bonds is usually necessary for the conversion of the proenzymes to the active enzymes; however, in some systems (e.g. procarboxypeptidase activation, YAMASAKI et al., 1963) this process involves the removal of a large part of the zymogen molecule. The immunological study of a given proenzyme and the immunochemical relationship to its affiliated enzyme will permit not only the characterization of both constituents as antigens, but can also give an indication of the structural and conformational changes, or of the "exposure", of regions, involved in the activation process.

Another important feature of many enzymes is the participation of a prosthetic group, such as a metal atom, in the construction of the active molecule. In most cases this group is essential for the enzymatic activity. Immunological studies on apoenzymes and the respective enzymes can point to the role which the prosthetic group, and the area contiguous to it in the enzyme molecule, play as antigenic determinants, or in stabilizing the native conformation. Comparisons might be drawn with the conversion of metmyoglobin to apomyoglobin (CRUMPTON, 1966). In this case antibodies specific towards haem-free apomyoglobin did react with metmyoglobin to yield a positive precipitin reaction, but the ferrihaem group could not be detected in the precipitate. These findings do not only serve as corroborating evidence for the existence of conformational differences between the two proteins, but also indicate that the combination with anti-apomyoglobin antibodies induced a conformational change in metmyoglobin.

There are other properties of enzymes which are suitable for immediate investigation by an immunological approach. These concern, on the one hand, the genetic basis of the multiple forms in which many enzymes are found, and on the other hand, the existence of allosteric enzymes. The occurrence of multiple forms of enzymes has been known for a long time and recently has been recognized as a general phenomenon. Most of these multiple molecular forms have been found by electrophoresis, but immunochemical methods are available and may play a role in the quantitation of the different forms and in detecting the structural differences between them. Furthermore, since we assume that it is a difference in the genes controlling the synthesis of these multiple forms which is the basis for their occurrence, we can generalize the phenomenon of multiple molecular forms to include genetically altered enzymes as well in this class of enzymes. The very interesting results concerning differences detected in single amino acid substitution in genetically altered alkaline phosphatase (COCKS and WILSON, 1969) may serve as an excellent illustration of this point.

It is not the purpose of this article to serve as an extensive survey of the literature. In the following pages I shall try to discuss the various problems mentioned above, and the impetus given to their exploration by the immunochemical studies of the pertinent enzymes in each case. The main emphasis, however, will be on the contribution of the many different antigenic determinants which are present on the enzyme surface to their overall immunological properties, and the consequent potential possessed by antibodies to enzymes for the identification and elucidation of such specificity determinants.

## 2. Inhibition and Enhancement of Enzyme Activity by Specific Antibodies

### a) Mechanism of Inhibition

The interaction between enzymes and their respective antibodies leads generally to a reduction in the enzyme activity. In some cases the enzyme is completely inhibited by the antibody, in others — partially and in a few cases no inhibition can be detected. A few exceptions to this phenomenon have also been reported, thus it has been shown that sometimes antibodies can even stimulate the activity of the enzyme, but this effect has been manifested mainly when poor substrates were used for the assay of catalytic activity or with mutant enzymes which by themselves have low catalytic activity (reviewed by CINADER, 1967). Several attempts have been made to elucidate the mechanism of the inhibition; however, it has not been found possible to arrive at a unified concept on the basis of results obtained with a large number of different enzymes (MARRACK, 1950; NAJJAR and FISHER, 1956; CINADER, 1957, 1967). Hence, regardless of whether a single mechanism can be ascribed to the reduction of the enzymic activity or whether each enzyme-antibody system represents a separate and unique problem, a number of factors are considered for the evaluation of the mechanism of inhibition for any particular enzyme which is investigated. The mechanism is then defined according to these factors, as will be illustrated by a few examples in the following:

a) The extent of inhibition of an enzyme by its antibodies is frequently related to the size of the substrate, as has been shown for several enzyme — anti-enzyme systems such as ribonuclease (BROWN et al., 1959; BRANSTER and CINADER, 1961), neuraminidase (FAZEKAS DE ST. GROTH, 1963), trypsin (ARNON and SCHECHTER, 1966) and papain (SHAPIRA and ARNON, 1967). These results led to the conclusion that the inhibition by antibodies is attributable mainly to steric hindrance. The effect is not necessarily due to the formation of antigen-antibody aggregates, as indicated, for example, by the finding that monovalent papain-produced anti-RNase fragments, which are capable of forming only soluble complexes, nevertheless inhibited the enzymatic activity of RNase on RNA more efficiently than on the small molecular weight substrate cytidine 2′, 3′-cyclic phosphate (CINADER and LAFFERTY, 1964).

b) Still, aggregate formation does contribute to steric hindrance, since it interferes with the access of substrate to the catalytic site. This has been demonstrated for several enzymes such as muscle glycogen phosphorylase (MICHAELIDES et al., 1964), carbamyl phosphate synthetase (MARSHALL and COHEN, 1961) and papain (SHAPIRA and ARNON, 1967). The additional inhibitory effect observed in these cases due to aggregate formation may reflect the relative positioning on the enzyme molecule of the antibody combining sites, namely the antigenic specificity determinants on the one hand, and the catalytic site on the other.

c) An additional factor which has been found to participate in the inhibition involves conformational changes imposed on the enzyme by its interaction with the antibody. The role of such changes in the mechanism of inhibition has been suggested by NAJJAR and FISHER (1956), and was implied also from the findings of SAMUELS (1963), who showed that the substrate can protect an enzyme like creatine kinase for example, from subsequent inactivation by its specific antibodies. Direct evidence for this effect has been provided by enzyme systems such as penicillinase, in which the interaction with the antibodies brought about enhancement of the enzymatic activity (POLLOCK et al., 1967). In this case it was shown that the activation is manifested mainly when the penicillinase is assayed on substrates which are poorly hydrolyzed by the enzyme. Consequently ZYK and CITRI (1965, 1968b) proposed that the effect of the antibody is to constrain the conformative changes imposed on the enzyme by its substrate, and therefore would result in the inhibition of hydrolysis of the more susceptible substrate, but in the apparent stimulation of the hydrolysis of substrates which are more resistant to the action of penicillinase. Stimulation of catalytic activity by the specific antibodies has been observed also with several other enzymes such as amylase (OKADA, 1963), ribonuclease (SUZUKI et al., 1969) and $\beta$-galactosidase (ROTMAN and CELADA, 1968). In the last case MESSERS and MELCHERS (1970), working with mutants' enzyme, proposed evidence that the activation which accompanies the interaction by the antibodies certainly involves conformational changes.

It can be concluded, therefore, that antibodies to enzymes affect the catalytic activity by steric hindrance and/or conformational changes, and that their mode of action is different for different enzymes.

### b) Inhibition — A Reflection of the Nature and Distribution of Antigenic Determinants

One of the features characterizing the inhibition of most enzymes by their respective antibodies is the residual catalytic activity persisting even in extreme antibody excess, an activity which is not reduced by the addition of more antibody. This effect has been observed with many systems, especially when low molecular weight substrates were used for the activity assay (reviewed by CINADER, 1957, 1963, 1967). This phenomenon can be interpreted at least in two ways. One possible explanation is that the antibodies inhibit according to a uniform mechanism; each enzyme molecule is partially inhibited, while retaining a residual enzymic activity after its combination with the antibody. Alternatively, the antibody population could be regarded as being inherently heterogeneous, consisting of species which differ in their inhibitory capacity. Undoubtedly, many antigenic determinants are present on the surface of each enzyme molecule and those are apt to give rise to heterogeneous antibody populations. There is no reason to assume that all or any of these antigenic determinants should include the catalytic site or the substrate-binding site of the molecule. On the other hand, if antibodies should exist whose specificity were directed towards groupings associated with the active center of the enzyme, their reaction with it would be expected to bring about inhibition of the enzymic activity, and the inhibitory capacity of such antibodies could indeed be higher than that of antibodies whose specificity is directed towards other regions of the antigen. The role of the antibody in the inhibition of the catalytic activity of an enzyme would then depend largely on its narrow specificity.

The validity of this second premise was demonstrated in studies with several enzyme systems by the actual separation of the antibodies into fractions that varied in their inhibitory capacity. Enzyme systems in which the reaction with antibody may result in enhancement of the catalytic activity are a case in point. For example, POLLOCK (1964) has shown that antisera against penicillinase contained both inhibiting and stimulating antibodies and he suggested a procedure for the enrichment of the serum in either of these activities (adsorption of the sera with small amounts of enzyme). Subsequently, ZYK and CITRI (1968a) fractionated the antibodies using Pollock's procedure, and succeeded in precipitating the inhibitory antibodies alone, leaving behind a supernatant solution which retained only the stimulatory antibodies. A similar approach, namely, the use of a small amount of enzyme for the limited selective precipitation of the antibodies, was also employed by FUCHS et al. (1969) in the staphylococcal nuclease system. In this case the authors fractionated the antibodies according to their mode of interaction with the enzyme in the presence or absence of $Ca^{++}$ and substrate analogues. Although the resultant two fractions were capable of inhibiting nuclease activity, they probably acted through different mechanisms, since only the inhibition by the fraction retained in the supernatant could be protected by $Ca^{++}$ and substrate analogues. It appears, therefore, that this is the fraction that interacts with determinants that are related to the substrate-binding site of the enzyme.

Selective fractionation of antibodies was also achieved in another system shown to contain activating antibodies. Thus, SUZUKI et al. (1969) succeeded in isolating one rather homogeneous fraction of stimulatory antibodies alongside ten other fractions of neutralizing antibodies from anti-ribonuclease, by column fractionation on DEAE-Sephadex. In contrast to the cases reported beforehand, the fractionation in this case was based, therefore, on differences in charge between the various antibody fractions. Antibodies to lactate dehydrogenase have also been fractionated, either by rate zonal ultracentrifugation in sucrose gradient, or by differential elution from a DEAE-cellulose column, into several distinct fractions with differing properties (NG and GREGORY, 1969). Some of these fractions gave very high titers in the passive hemagglutination reaction with the enzyme but caused little or no inhibition of its catalytic activity; other fractions were capable of inhibiting the enzyme but caused little or no hemagglutination. The non-inhibitory antibodies could protect the enzyme from inactivation by the inhibitory antibodies.

All the examples cited above clearly indicate that the total anti-enzyme population indeed consists of antibody species which differ in their inhibitory properties, and that these may be separated by suitable means. Evidence for the assumption that the inhibitory capacity of the antibodies is indeed dependent on their narrow specificity, namely on the antigenic determinants of the enzyme with which they combine, was given by the system papain — anti-papin (ARNON and SHAPIRA, 1967). In this case, antibodies with different inhibitory capacities were fractionated on the basis of their ability to cross-react with a related enzyme, chymopapain — which presumably contains similar antigenic determinants. Cross-reaction between isofunctional enzymes, either those isolated from different species or those obtained from different organs of the same species, is a widespread phenomenon (see Chapter 3). In most of these cases the cross-reaction is accompanied by cross-inhibition of the enzymes by the respective cross-reacting antibodies. Such cross-inhibition was observed not only between equifunctional enzymes of different origins, but also between different enzymes having similar active sites such as trypsin and chymotrypsin (ARNON and SCHECHTER, 1966). In the case of papain and chymopapain the two enzymes cross-precipitate with each other and are also cross-inhibited by their respective antisera. It could, therefore, be assumed that the regions in the molecule which the two enzymes have in common include those antigenic determinants whose interaction with the antibodies is responsible for the decrease in catalytic activity of both enzymes. The fraction of the antibodies in anti-papain serum that cross-reacted with chymopapain was consequently isolated on a chymopapain immunoadsorbent, and was indeed shown to possess high inhibitory capacity, much higher than that of the total antibody preparation. On the other hand, the antibodies that could not bind to the chymopapain immunoadsorbent were hardly inhibitory at all (Fig. 1). Hence, the partial inhibition of papain by the totality of its antibodies may be regarded as an over-all value expressing the probability of the interaction of the enzyme with the various antibody species which constitute the heterogeneous antibody population.

The antibodies which comprise the inhibitory fraction are specific for those antigenic determinants on the papain molecule which are present on chymopapain as well, and as expected they reacted equally well with the two enzymes, both in forming antigen-antibody complexes and in inhibiting their enzymic activity. Moreover, antibodies with identical properties were subsequently isolated also from anti-chymopapain serum by employing a parallel procedure (ARNON and SHAPIRA, 1968). The antigenic determinants common to the two related enzymes therefore served as haptenic groups for the selection of antibodies with the same determinant specificity elicited by two different antigens; the handle to prove this effect was the inactivation of catalytic activity.

Another system in which the inhibition of catalytic activity led to the fractionation of antibodies according to their narrow specificity is lysozyme —

anti-lysozyme. The existence of non-inhibitory antibodies in this system was indicated in early experiments (SHINKA et al., 1962) by the formation of some antigen-antibody complexes which were enzymatically active. In later studies, it was demonstrated that antibody fractions with different inhibitory properties could actually be separated from anti-lysozyme serum. This was achieved by two different procedures. One approach was based on the availability of methods for the isolation of immunologically active fragments of lysozyme containing distinct antigenic determinants (SHINKA et al., 1967; FUJIO et al.,

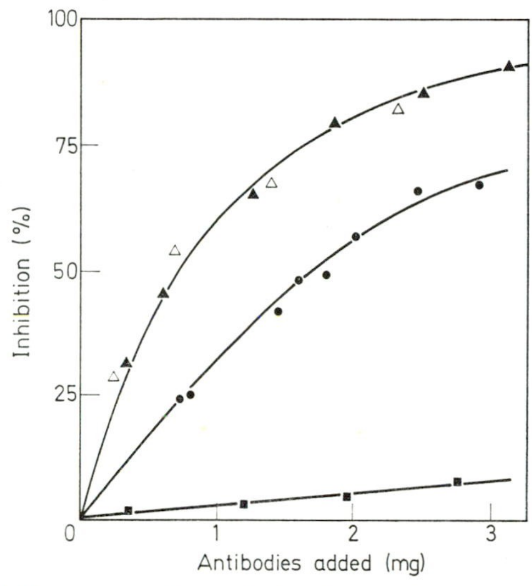


Fig. 1. Inhibition of the enzymatic activity of papain (50 μg) on benzoyl-L-arginine ethyl ester by the total anti-papain antibodies preparation (●) and by the two selectively fractionated species, *i.e.* the fraction that was isolated on chymopapain immunoadsorbent (▲) and the antibodies that could not bind to chymopapain immunoadsorbent (■). The open triangles (△) indicate the inhibition of chymopapain (50 μg) by the fraction of anti-papain antibodies that was isolated on chymopapain immunoadsorbent. Source: ARNON and SHAPIRA (1967)

1968a, 1968b; ARNON, 1968). Two such fragments were isolated, one peptide was derived from the portion $Gln^{57}$ to $Ala^{107}$, and the other from the amino and carboxy terminal regions of hen egg-white lysozyme. Utilization of these fragments for the preparation of immunoadsorbents led to the isolation of antibody fractions with different capacities to inhibit the catalytic activity of lysozyme (IMANISHI et al., 1968; ARNON, 1968). However, neither of these selected antibody fractions had higher inhibitory capacity than that possessed by the total antibody population, or by a mixture of the two separated fractions (Fig. 2). This was observed when the enzyme activity was assayed with both high and low molecular weight substrates. It appears, therefore, that the antibodies specific to these two antigenic determinants of lysozyme do not play a direct, decisive role in the neutralization of the enzyme by its antibodies, a finding which is not surprising in view of the knowledge that the amino acid sequences of which these fragments consist are not involved in the catalytic function of the enzyme (PHILLIPS, 1967).

A highly inhibitory antibody fraction from anti-lysozyme serum was recently isolated by means of a different approach: based on the findings of FELLENBERG and LEVINE (1967) that a small molecular weight lysozyme inhibitor, tri-N-acetyl glucosamine, partially inhibited the serological activity of lysozyme, IMANISHI et al. (1969) used the same inhibitor for the dissociation of lysozyme — anti-lysozyme complexes. The antibodies fractionated in that manner (comprising 7—8% of the total precipitating antibody) were

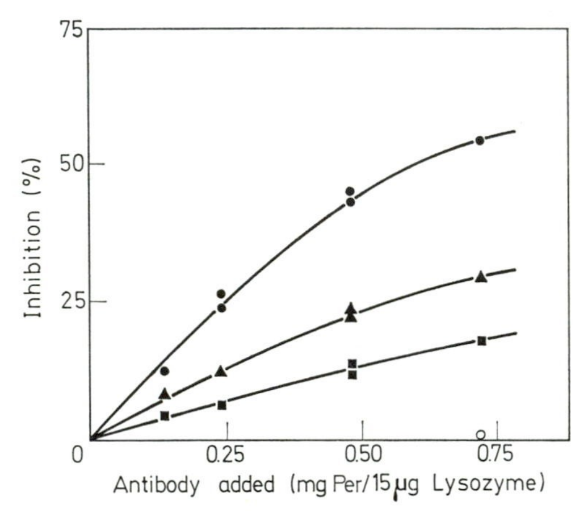


Fig. 2. Inhibition of the catalytic activity of lysozyme on penta-N-acetyl glucosamine by the various antibody species: ● -total antilysozyme antibodies; ■ -the fraction of antibodies isolated on an immunoadsorbent containg a lysozyme fragment (residues 60—83 and 91—108); ▲ -the antibodies which could not bind to the above immunoadsorbent. Source: ARNON (1968)

efficient inhibitors, even as regards the lysozyme activity on small molecular weight substrate. Although this antibody fraction was still precipitable with lysozyme, an effect implying that these antibodies are not specific to the substrate-binding site exclusively, a selective fractionation based on inhibitory capacity was indeed achieved in this case, resulting in antibodies toward a limited number of antigenic sites.

In conclusion, the inhibitory properties of the antibodies are dependent on the determinants towards which they are specific. In several cases, methods have been devised for the isolation of antibodies specific to determinants which are related to the catalytically active site. Those antibody fractions carry the inhibitory capacity.

### 3. Inhibition of Catalytic Activity — A Probe into the Study of Evolution of Enzymes

In general, the process of evolution has been in the direction of greater diversity of the enzymes both in chemical structure and in biological specificity. Whereas the latter leads to the occurrence of homologous enzymes with different specificities such as trypsin and chymotrypsin (WALSH and NEURATH, 1964), the former led to the existence of equifunctional enzymes in many species or in various organs of the same species. The relationships between such phylogenetically homologous enzymes involve not only chemical similarity, but conformational homology as well, for which immunological cross-reaction may be taken as corroborating evidence. Isofunctional enzymes may bear greater similarity in regions of the molecule which participate in the catalytic activity. In these cases one may expect considerable inhibition of the catalytic activity by cross-reacting antibodies. And, just as immunological cross-reaction indicates similarity between antigenic determinants, the cross-inhibition may be taken to imply similarity of the structure of the catalytic center or regions related to it on the enzymes' surface.

The studies reported in the literature are concerned with the two types of relationships: on the one hand, the comparison of enzymes that originate from different organs of the same species, and on the other hand, studies comparing isofunctional enzymes from different species. An interesting observation in several investigations in which a particular enzyme isolated from different organs was subjected to such a comparative study, is the finding of a large extent of cross-reaction, implying almost complete identity between enzymes originating in some organs, contrasted with the lack of any cross-reaction with the enzyme derived from another organ of the same species. Amylases from hog pancreas and saliva, for example, were reported to be very similar to each other and distinctly different from hog liver amylase (MCGEACHIN and REYNOLDS, 1960). Similarly, fructose-1,6-diphosphatases from liver and kidney were found to be identical according to the criteria of both gel diffusion and inhibition by antibodies, but were completely different from the muscle enzyme which did not precipitate with the antibodies to the liver enzyme, nor was it inhibited by them (ENSER et al., 1969). Close similarity was also reported between lactic dehydrogenases from human heart, kidney, prostate, brain and erythrocytes but no relationship with the enzyme isolated from human liver or skeletal muscle (NISSELBAUM and BODANSKY, 1961). Human alkaline phosphatase from different organs such as liver, placenta or kidney were found to be distinctly different from each other, not allowing any immunological cross-reaction. On the other hand, cathepsins D isolated from liver, spleen, heart, kidney, testis, brain and limb bones were all found to be identical according to their reaction with specific antiserum to liver cathepsin D (WESTON, 1969). It appears, therefore, that enzymes at the same evolutionary level are probably very similar or even identical; enzymes from the same species that do not relate immunologically may constitute different proteins altogether, although they catalyse the same reaction.

Enzymes isolated from different species in most cases do cross-react with each other, but they exhibit much more gradual differences and similarities than enzymes from the same species. The extent of cross-reaction and cross-inhibition by their antibodies seems to be dependent on the phylogenetic distance between the species compared and on the similarity in their amino acid sequence. A few examples will illustrate this point: lysozymes from 16 species of birds were examined for their reactivity in the complement fixation test with anti-hen egg-white lysozyme (ARNHEIM and WILSON, 1967). The strongest cross-reaction was obtained with quail lysozyme, while pheasants were at the weaker end of the reactivity series. These results were unexpected since, according to zoological evidence, pheasant is closer to hen; however, a subsequent comparison of the amino acid sequences of these lysozymes revealed only one or two amino acid interchanges between hen and quail lysozymes, compared to seven interchanges in the pheasant enzyme (ARNHEIM et al., 1969) thus indicating that the amino acid sequence is a more decisive factor than zoological relationship. In another study MARON et al. (1970a) have shown the extent of relatedness between several bird and human lysozymes. According to their observations, guinea-hen lysozyme is closer to hen lysozyme than the duck protein, and the latter may be separated into two chromatographic fractions which differ in their reactivity with anti-hen lysozyme, a finding which is consistent with the known replacement of one glycine residue by arginine in the less reactive species (JOLLÈS et al., 1967). Among the human proteins — milk lysozyme gave a definite and reproducible reaction, in contrast to enzyme from normal leucocytes, suggesting that the latter has even less in common with hen egg-white lysozyme than the human milk lysozyme.

Another example for such gradual divergence was offered by the study of catalases (SZEINBERG et al., 1969). Antibodies to human erythrocyte catalase reacted with catalases from various species, but according to the extent of cross-reaction the animals could be classified into four groups within which identical reaction was observed: 1) man, Rhesus monkey, Vervet monkey; 2) horse, donkey, guinea pig; 3) goat, sheep, calf; 4) dog. The order of cross-reaction in this case relates to the evolutionary pattern.

Studies with trypsins from various species were also aimed at the elucidation of the evolutionary pattern. The cross-reactions between four different trypsins (originating from both higher mammals and lower vertebrates), and the respective antisera to all of them were compared. According to the relative immunological cross-reactions and inhibitory effects caused by the antibodies, conclusions could be reached about the extent of relatedness among these different trypsins and about the order in which these enzymes developed during evolution. The order of "similarity" between them, according to the immunological criteria, was established to be bovine > porcine > dogfish > starfish, indicating that trypsin existed millions of years before the appearance of man (ARNON and NEURATH, 1969).

The enzyme most extensively studied in this respect is undoubtedly cytochrome *c* (MARGOLIASH et al., 1970). Precipitating antisera against human, monkey and horse cytochromes *c* were prepared, and compared for their cross-reactivity with cytochromes *c* from 25 different species. It was found that cytochromes *c* from different species which have identical amino acid sequences were immunologically indistinguishable. On the other hand, in those cytochromes *c* which differ by a single amino acid residue, the immunological distinction may vary from very minor to major. When the difference involves several residues, there is a rough correlation between the number of differences in primary structure among the various cytochromes *c* and the homologous antigen, and the extent of their immunological cross-reactivity. This subject will be discussed in more detail in the next chapter which will describe the identification of antigenic determinants of cytochrome *c*.

In some cases the lack of immunological cross-reaction or cross-inhibition by the antibodies may also have a bearing on phylogeny. A case in point is the relationship between hen egg-white lysozyme and bovine α-lactalbumin. These two proteins have been shown to possess similar structural features (BREW and CAMPBELL, 1967): they have similar molecular weights, a similar number of disulfide bonds, and closely related amino acid sequences in which the four disulfide bonds are in identical positions (BREW et al., 1967). It was suggested that the two proteins possess a similar molecular conformation, and indeed it has been possible to fit the side chains of bovine α-lactalbumin to a wire skeletal model of the polypeptide backbone of lysozyme (BROWNE et al., 1969) and to generate a molecule which retains the major structural features of lysozyme. The explanation for the existence of structural homology was given in terms of evolutionary relationships, since α-lactalbumin participates in the catalytic reaction of lactose synthetase (BREW et al., 1968) which is inversely related to the reaction catalyzed by lysozyme. Notwithstanding this aggregate of evidence for structural homology, no immunological cross-reaction has been observed between lysozyme and α-lactalbumin either in the precipitin reaction (ATASSI et al., 1970) or when tested by six additional immunological techniques of high sensitivity (ARNON and MARON, 1970); nor were antibodies to lactalbumin capable of inhibiting the catalytic activity of lysozyme. It must be concluded, therefore, that conformational or functional homologies among different proteins are not necessarily paralleled by the presence of antigenic determinants of similar specificity, whereas phylogenetic variations of a single enzyme show antigenic similarity and provide a powerful tool for tracing the process of evolution.

### 4. Identification of Antigenic Determinants

An antigen can provoke formation of antibodies against many different determinants, each determinant being defined as that part of the molecule which reacts immunospecifically with the combining site of the antibody. Those may consist of "sequential" determinants — which are due to stretches of amino acid sequences in the protein and "conformational" determinants which result from the steric conformation of the macromolecule (SELA et al., 1967). Studies on the antigenic structure of protein molecules (reviewed by KAMINSKI, 1965) have indicated that fibrilar proteins, such as silk fibroin, contain many sequential determinants — on enzymic cleavage they yield small peptides which are immunologically active (CEBRA, 1961) — whereas globular proteins contain mainly, but not exclusively, conformational determinants.

Attempts to identify antigenic determinants have been carried out with several proteins and enzymes, utilizing various approaches. One of the approaches has also been applied to a number of globular proteins which are not enzymatically active, such as bovine serum albumin (PORTER, 1957), human serum albumin (LAPRESLE and DARIEUX, 1954) sperm whale myoglobin (CRUMPTON, 1967; CRUMPTON and WILKINSON, 1965; GIVAS *et al.*, 1967; ATASSI and SAPLIN, 1968) and tobacco mosaic virus protein (ANDERER, 1963; BENJAMINI et al., 1964, 1965). It involves the fractionation of fragments obtained by limited proteolytic digestion of the enzyme in question, and the screening of the resultant fractions for immunologically active components. These components, which by definition embody antigenic determinants, are subsequently analyzed and defined. This technique was employed in the study of the antigenic structure of a number of enzymes, as illustrated by the following examples.

In the case of hen egg-white lysozyme (SHINKA et al., 1967; ARNON, 1968; FUJIO et al., 1968a, 1968b), this approach enabled the localization of two independent antigenic determinants on the molecule. One of these immunologically active fragments consisted of two peptides derived from the $NH_2$-terminus (residues 1—27) and the COOH-terminus (residues 122—129) of lysozyme, linked together by a single disulfide bond. This peptide bound to anti-lysozyme antibodies with an average affinity constant of $1.75 \times 10^5$ and the percentage of antibodies directed towards it was evaluated at 47% (FUJIO et al., 1968a). The second immunologically active component isolated was, as already mentioned, a large fragment derived from the region located between residues 57 and 107 of the lysozyme sequence (CANFIELD and LIU, 1965; SHINKA et al., 1967; ARNON, 1968; FUJIO et al., 1968b). This peptide, which contains two disulfide bridges, was also capable of binding to anti-lysozyme antibodies and in that way to interfere both with their precipitation with lysozyme and with their inhibition of its catalytic activity. The antibody fraction specific towards this peptide (amounting to 30% of the total antibodies) inhibited lysozyme activity. Each of these two immunologically active fragments is quite large, and probably contains more than a single antigenic determinant. It was shown later that the last mentioned fragment can yield a smaller peptide which still retains immunological activity (ARNON and SELA, 1968). This fragment, consisting of the amino acid sequence 60—83 and containing one intrachain disulfide bond, was denoted "loop" (Fig. 3). Antibodies to this region only, prepared either by selective isolation from anti-lysozyme serum on a "loop" immunoadsorbent, or by immunization with a synthetic "loop" conjugate, showed, as expected, less heterogeneity than the total anti-lysozyme antibody population (MARON et al., 1970b). Moreover, these antibodies were directed toward a conformation-dependent determinant, as indicated by their capability to distinguish between the "loop" and its open peptide chain, and could recognize the "loop" structure in native lysozyme. Consequently, they reacted with the intact enzyme even though they could not precipitate with it or inhibit its catalytic activity. In view of the knowledge

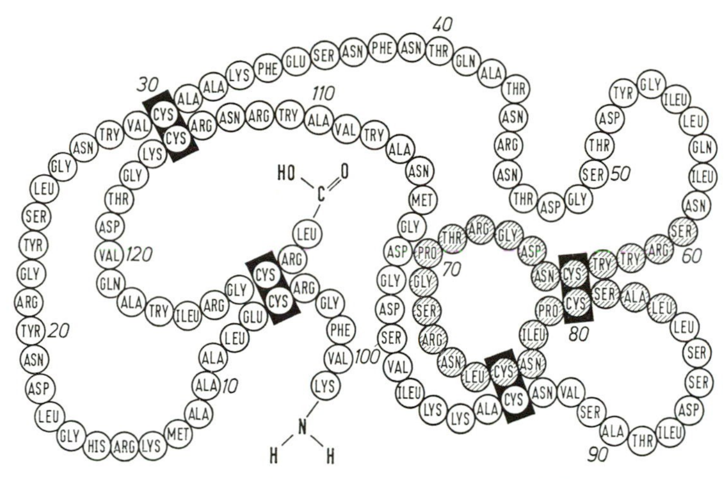


Fig. 3. Amino acid sequence of hen egg-white lysozyme. The region of the "loop" peptide is shaded. Source: Arnon and Sela (1969)

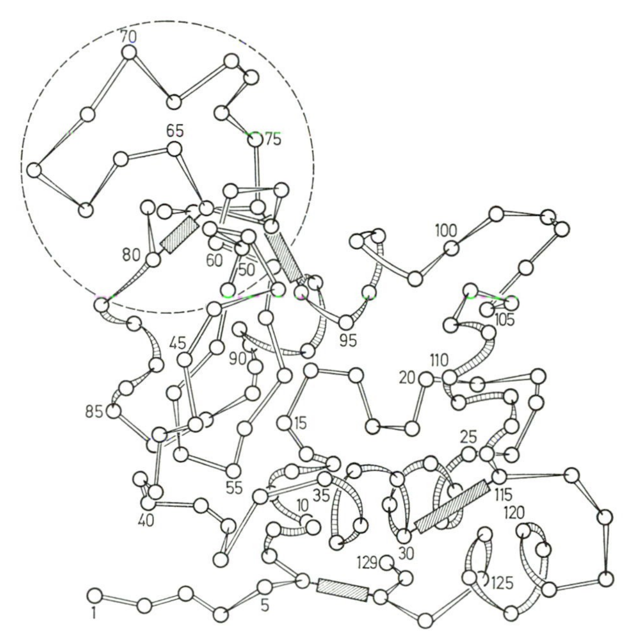


Fig. 4. Schematic drawing of the main chain conformation of hen egg-white lysozyme. The area encompassing the "loop" peptide is encircled. Source: Blake et al. (1965)

of the three-dimensional structure of lysozyme, such information identifies in precise molecular terms the locus on the enzyme that encompasses an antigenic determinant (Fig. 4).

It is of interest that in studies with completely reduced and carboxymethylated lysozyme — a derivative which does not cross-react at all with native lysozyme — a peptide from *the same* region in the molecule (residues

74—96) was found to possess immunological activity — it was capable of inhibiting the homologous antigen-antibody reaction of the open chain lysozyme (Gerwing and Thompson, 1968). These findings indicate that this region of lysozyme contains both conformational and sequential antigenic determinants.

Studies with oxidized ribonuclease — which does not cross-react with native ribonuclease — also pinpointed linear sequences that encompass antigenic determinants. Thus two peptides isolated from a proteolytic digest of the protein, comprising residues 38—61 and 105—124, were found capable of inhibiting the precipitation and complement fixation of the oxidized ribonuclease with its antibody (Brown, 1962). Recently it was shown that the peptide 105—124 can bind to antibodies against oxidized ribonuclease with an association constant of $3 \times 10^6$ and a heterogeneity index of 0.98. This indicates that the peptide is bound to a rather homogeneous fraction of the antibodies, with relatively high affinity, which, according to the calculation, represents about 50% of all the precipitable antibodies to this antigen (Isagholian and Brown, 1970). It is of interest to note that this peptide contains two out of the four proline residues of ribonuclease.

A similar technique, namely the use of a digest of the enzyme for the isolation of immunologically active fragments, was used in the identification of antigenic determinants in another enzyme — staphylococcal nuclease. Fuchs et al. (1969) demonstrated that this enzyme elicits the formation of antibodies capable of inhibiting its catalytic activity. Recently, Omenn et al. (1970a) prepared peptide fragments of nuclease, with known amino acid sequence, which were able to interfere with the nuclease — anti-nuclease reaction. The peptides were prepared by cyanogen bromide cleavage, by limited tryptic digestion and by solid-phase synthesis. One set of peptides, representing overlapping sequences from the carboxy-terminal portion and another set from the amino-terminal portion of the enzyme were immunologically active. According to those data the presence of antigenic determinants could be localized to the linear sequences 127—149 and 18—47 (possibly 18—26) of the nuclease molecule. In addition to these determinants, however, the existence of conformation-dependent determinants as well was inferred from the enhanced binding of the peptides to the antibodies when incorporated into a non-covalent, enzymatically-active complex, as compared to the binding of the peptides as such. Thus, although nuclease has a low helix content, and lacks any disulfide structure (Taniuchi and Anfinsen, 1968; Cuatrecases et al., 1968a), its structural conformation influences its antigenic determinants.

A somewhat different approach was employed for the identification of antigenic determinants on another enzyme of known tertiary structure, namely chymotrypsin (Sigler et al., 1968). In a study by Sanders et al. (1970) an attempt was made to verify the existence of conformational homology between the presumably homologous enzymes, trypsin and chymotrypsin, by demonstrating the presence of similar antigenic determinants on both enzymes. For this purpose only immunologically cross-reacting fragments, i.e. fragments which would bind to anti-trypsin antibodies, were isolated from a peptic digest of chymotrypsin. Two peptide fragments, each containing two peptide chains linked by a disulfide bond, were isolated and identified, and each was shown to be capable of reacting with both anti-trypsin and anti-chymotrypsin; one of them contained the active serine of the enzyme, a residue which is

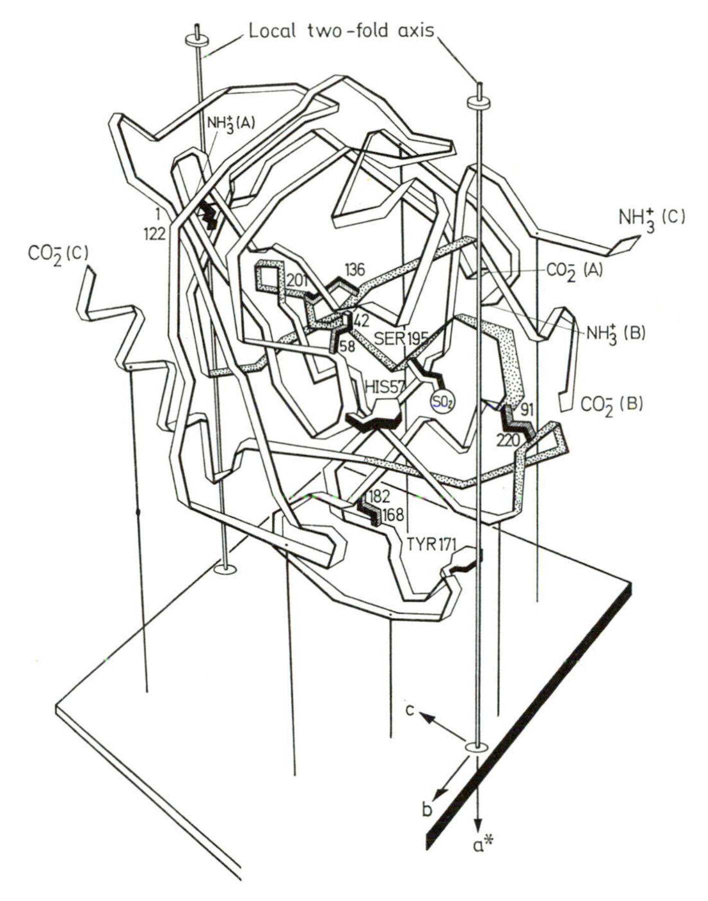


Fig. 5. Drawing of a model of α-chymotrypsin three-dimensional structure. Shaded area denotes the immunologically active peptides that were isolated by their binding the anti-trypsin antibodies. Source: SANDERS et al. (1970)

a common feature present not only in trypsin and chymotrypsin, but also in a group of enzymes classified as "serine proteases". According to the three-dimensional model of chymotrypsin the two antigenically active fragments which were isolated occupy corners of the polypeptide chain (Fig. 5) allowing the prediction that the corresponding antigenic determinants on trypsin might occupy similar positions in its spatial structure. In this case, the search for defined antigenic determinants on enzymes proved to be helpful in the establishment of conformational relationships in structures which are homologous in the chemical sense.

Upon inspection and comparison of the fragments which encompass the antigenic determinants of the various enzymes described till now, we arrive at an interesting conclusion — most of these fragments contain proline residue in a much higher amount than could be expected according to the proline content in the protein. Thus, the fragments comprising the common antigenic determinants of trypsin and chymotrypsin contain two homologous prolyl residues; the immunologically reactive "loop" region of lysozyme embodies both prolyl residues which are the only ones present in the molecule, and the same segment consists of an immunodominant region in the reduced open-chain derivative of lysozyme. Likewise, one of the two immunologically active regions in staphylococcal nuclease (residues 18—47) contains three out of the six proline residues of the enzyme, and similarly in oxidized ribonuclease the two immunologically active peptides contain three out of the four prolines of the molecule. These separate observations, corroborated also by findings in other proteins such as spermwhale myoglobin (ATASSI and SAPLIN, 1968) or tobacco mosaic virus (BENJAMINI et al., 1965) raise the question of whether a proline residue can represent a feature on the surface of protein molecules that can constitute a recognition point for the immune system, and thus serve as a dominant element in the structure of antigenic determinants.

It must be remembered, however, that the approach described till now, namely the search for immunologically active fragments, cannot be expected to yield *all* the antigenic determinants on any protein, because of the fact that some of the determinants are undoubtedly destroyed by denaturation and by the digestion process. The possible participation of other regions of the molecule in antigenic specificity determinants, besides those that were identified in the various cases, cannot therefore be excluded. Furthermore, the conformation of an isolated peptide is not necessarily the same as that occupied by the same peptide in the intact enzyme. Although the antibody may in some case influence or stabilize the "native" conformation of the peptide, the possibility exists that a peptide originating from an antigenically active region, will escape notice.

A completely different course towards the elucidation of antigenic determinants on enzymes was taken in the studies with the cytochromes *c* (NISONOFF et al., 1970). These authors did not look for immunological activity on fragments of the molecule, but rather utilized the cross-reaction between intact cytochromes *c* of various sources. They anticipated that when minimal differences exist in the amino acid sequences of the proteins under comparison, and those are exactly delineated, the results may be used to identify and localize some of the antigenic determinants. For example, it was found that 30—40% of the antibodies against human cytochrome *c* failed to react with *Macaca mulata* cytochrome *c*, which differs from the human protein only at residue 58, where it has a threonyl residue instead of isoleucyl residue (Fig. 6). On the other hand, the antibodies which failed to react with the monkey's protein, in addition to their reaction with human cytochrome *c*, reacted only with the kangaroo's protein. This is the only cytochrome *c* of all those studied which has an isoleucyl residue at position 58, similar to the human protein (Table 1). Interestingly, in constrast with the behaviour of anti-human cytochrome *c*, antibodies to *M. mulata* cytochrome *c* reacted identically with the simian and human proteins, thus indicating that the occurrence of an immunogenic determinant is related to the presence of the hydrophobic isoleucine, whereas in the absence of isoleucine this region is immunologically inert. Quantitative analysis indicates that the "isoleucine site" is antigenically identical in the human and the kangaroo proteins, two

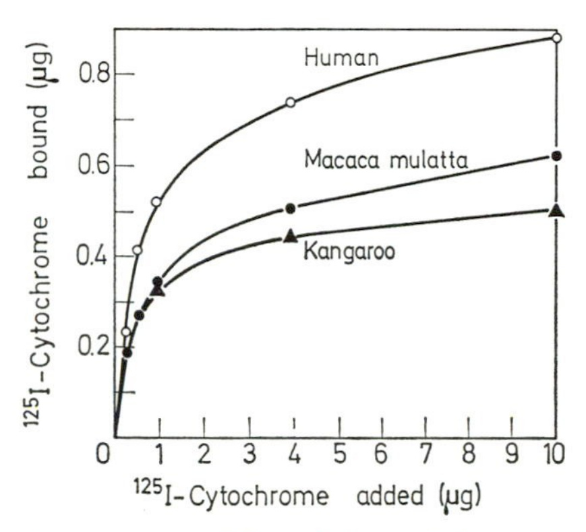


Fig. 6. Binding of $^{125}$I-labeled human, *M. mulatta* and kangaroo cytochrome *c* by anti-human cytochrome *c*. Each tube contained 0.3 ml of 1:25 diluted antiserum. Source: NISONOFF et al. (1970)

Table 1. *Antibodies Specific to Human and Kangaroo Cytochrome c*

Displacement of $^{125}$I-labelled human cytochrome *c* by various unlabelled cytochromes *c* from anti-human cytochrome *c* previously adsorbed with an excess of cytochrome *c* of *M. mulatta*[a].

| Competing unlabelled cytochrome *c* | % Inhibition of binding |
|---|---|
| *M. mulatta* | 28 |
| Human | 96 |
| Kangaroo | 90 |
| Rabbit | 16 |
| Cow | 11 |
| Horse | 8 |
| Dog | 13 |
| Chicken | 13 |
| Turkey | 14 |
| Pekin duck | 8 |
| Tuna | 12 |
| Screw worm fly | 1 |

[a] Binding measurements were carried out by the modified Farr technique with $^{125}$I-labelled human cytochrome *c*. The percent inhibition of binding is given relative to the amount of $^{125}$I-cytochrome *c* bound to the adsorbed antiserum in the absence of competing proteins. Source: NISONOFF et al. (1970).

species that are not closely related. The appearance of this "isoleucine site" and its importance in immunogenicity emphasize the difficulties attendant upon attempts to analyze evolutionary pathways through antigenic analysis alone. However, the identification of antigenic determinants, combined with the elucidation of primary and tertiary structures, should eventually permit the mapping of antigenic regions in enzymes and in proteins in general.

Studies using a parallel but somewhat different approach have also been carried out with lysozymes. In this case, as in that of cytochrome *c*, enzymes isolated from different species cross-reacted with each other's antisera to a large

Table 2. *Relatedness of Various Lysozymes*

Inhibition[a] by various lysozymes of the inactivation of modified bacteriophages with anti-(hen egg-white) lysozyme.

| Lysozyme sources | Modified bacteriophage preparation | |
|---|---|---|
| | Lysozyme-$T_4$ | "Loop"-$T_4$ |
| Hen egg-white | $1.5 \times 10^{-6}$ | $1.7 \times 10^{-5}$ |
| Guinea hen egg-white | $6.3 \times 10^{-6}$ | $2.3 \times 10^{-5}$ |
| Duck egg-white II | $5.7 \times 10^{-4}$ | $1.4 \times 10^{-2}$ |
| Duck egg-white III | $5.7 \times 10^{-4}$ | —[b] |
| Human milk | $2.1 \times 10^{-3}$ | —[b] |
| Normal human leucocytes | —[b] | —[b] |

[a] The numbers in the table indicate the concentration (mg per sample) of each lysozyme which brought about 50% inhibition of the bacteriophage inactivation by the antibodies.

[b] No inhibition was observed with an inhibitor concentration as high as 0.15 mg per sample. Source: MARON et al. (1970).

extent (FUJIO et al., 1962; ARNHEIM and WILSON, 1967). Recently MARON et al. (1970a) investigated this cross-reaction in terms of a selective comparison of certain regions in the molecule. These authors employed the technique of inactivation by antibodies of modified bacteriophage (HAIMOVICH et al., 1970). By comparing the capacity of various bird and human lysozymes to interfere with the inactivation of lysozyme-bacteriophage conjugate or "loop"-modified bacteriophage by antibodies against hen egg-white lysozyme, the extent of relatedness of these lysozymes or their "loop" region to hen lysozyme could be estimated (MARON et al., 1970a). Thus it was observed that guinea hen lysozyme is only four times less efficient as an inhibitor than hen lysozyme in the lysozyme-phage system, whereas both have the same inhibitory capacity in the "loop"-phage system (Table 2). This implies that any differences in the "loop" region of these two proteins are not reflected immunologically. On the other hand, two duck lysozymes are indistinguishable in the lysozyme-phage system but differ completely in the "loop"-phage system. This immunological difference may be due to the known replacement of the glycine residue in position 71 of hen and duck lysozyme II with an arginine residue in duck lysozyme III (JOLLÈS et al., 1967).

In other studies, an attempt was made to compare the antigenic determinants of hen egg-white and turkey egg-white lysozymes. These proteins, which differ from each other by seven amino acid replacements, were found to cross-react with their corresponding antisera almost fully (SCIBIENSKI et al., 1970), indicating a large extent of similarity in their antigenic structure. Nevertheless, antibodies specific only to turkey lysozyme could be elicited in rabbits provided that they were previously made tolerant to hen's lysozyme (STRATTON et al., 1970). It seems that in this system, the seven amino acid residues which are different in the two enzymes, or even only some of them, correlate with at least one antigenic determinant on the turkey's lysozyme.

In summary, the study of molecular fragments of enzymes or the comparison of the immunological reactivity of homologous enzymes, has made it possible to determine and identify amino acid residues or groups of such residues which partake in the antigenic determinants of several enzyme molecules.

## 5. Structure-Function Correlation

Immunological studies of enzymes aimed at the elucidation of structure-function relationships have been carried out along two main avenues: on the one hand, various workers have studied effects of conformational changes brought about by either partial or complete denaturation of the enzyme or by unfolding of the polypeptide chain, on both the enzymic and the immunological properties. On the other hand, the specific modification of particular side chains in the molecules has been employed in an effort to identify those residues which partake in the enzymic activity or in the antigenic make-up of the molecule.

### a) Effects of Conformational Changes

Complete unfolding of the polypeptide chain in most cases brings about the loss of enzymic activity in parallel with the elimination of the capacity to interact with antibodies to the native protein. This has been demonstrated for several enzymes: performic acid-oxidized ribonuclease (Ribox) is enzymatically inactive and it does not cross-react at all with antibodies to the native enzyme (BROWN et al., 1959); similarly, completely reduced and phosphorothioated ribonuclease reacts very poorly with antibodies to the native antigen (NEUMANN et al., 1967). Completely reduced and carboxymethylated lysozyme also lacks both catalytic activity and the capacity to react with antibodies to the native enzyme (GERWING and THOMPSON, 1968; YOUNG and LEUNG, 1969). In both of the above instances of the completely unfolded polypeptide chain, namely the oxidized ribonuclease and the reduced and carboxymethylated lysozyme, the derivatives are capable of eliciting antibody formation, but these antibodies are specific to the unfolded peptide chain and do not cross-react at all with the respective native enzyme (BROWN, 1962; GERWING and THOMPSON, 1968). Papain is another example in which complete reduction and carboxymethylation lead to the elimination of the capacity to react with antibodies to the native enzyme (SHAPIRA and ARNON, 1969), and a similar phenomenon has been reported for pepsin (GERSTEIN et al., 1964) and trypsin (ARNON and NEURATH, 1970).

Partial reduction of disulfide bonds, on the other hand, has yielded derivatives which retain, to varying extents, both catalytic activity and the ability to react with antibodies to the native enzyme. In the case of ribonuclease, the steric conformation of a derivative in which two disulfide bonds had been reduced was shown to be looser than that of the native enzyme, as evidenced by its digestibility by trypsin, and yet the catalytic activity was not impaired at all and the antigenic specificity determinants were not destroyed (NEUMANN et al., 1967). Partial reduction of pepsin also resulted in a preparation which retained the capacity to react with antibodies to the native enzyme (GERSTEIN et al., 1964). Thus, with only one disulfide bond reduced, pepsin possessed immunological properties identical with those of the native enzyme; pepsin with two bridges reduced lost its capacity to react with the antibody directly but maintained the ability to inhibit the pepsin-antipepsin system. The reduction of all three disulfide bridges, as already mentioned, completely suppressed antigenic activity. Similarly, the cleavage of one disulfide bond in papain yielded a derivative which cross-precipitated 60% of the antibodies to native papain, in contrast with the non-reactive completely unfolded enzyme (SHAPIRA and ARNON, 1969).

The effect of another type of structural modification was tested by BONAVIDA et al. (1969) who introduced local conformational changes in lysozyme by cleavage with cyanogen bromide, while maintaining the disulfide bonds of the enzyme intact. They observed that both the catalytic activity and the susceptibility to trypsin digestion were considerably altered as a result of this treatment, whereas the immunological activity was affected to a lesser extent—although it had lost some antigenic determinants, the cyanogen bromide-treated lysozyme retained its capacity to form a precipitate with anti-lysozyme serum, and to inhibit to the extent of 70% the binding of labeled native lysozyme to its antibodies. It appears, therefore, that limited changes in the steric conformation of an enzyme are not always paralleled by a drastic decrease in its immunological reactivity, an effect which is always observed on total unfolding of the polypeptide chain.

Denaturation by physical means, like heat denaturation, high urea concentrations, or changes of pH usually brings about the loss of catalytic activity. Studies with several enzymes have indicated that under these conditions the reaction of the enzyme with the specific antibodies induced stabilizing effects, implying that the denaturing conditions did not prevent the enzyme-antibody interaction. For example, NAJJAR and FISHER (1956) mentioned that yeast alcohol dehydrogenase, which is ordinarily inactivated within several minutes at 4°, is stabilized by the homologous antibody and can then be maintained in the active form for months at that temperature. KAPLAN and WHITE (1963) also reported that the thermal stability of the antigen-antibody complex of lactic dehydrogenase depends on the amount of antibody present. Similarly, ZYK and CITRI (1968a, b) observed that complexing with the antibodies increased the stability of penicillinase both to heat and to variation of pH. The stabilization in this case was brought about either by antibodies which inhibited the enzymatic activity or by antibodies that stimulated activity. Another example is acetylcholine esterase, an enzyme which is not inhibited by its specific antibodies, but still shows higher heat stability in a complex with the antibody than as the free enzyme (MICHAELI et al., 1969a). This stabilization takes place both in enzyme-antibody precipitates and in soluble complexes. Moreover, even enzyme that had been previously denatured by heating could be reactivated by complexing with the antibody (MICHAELI et al., 1969b). This finding demonstrates that at least in this case, the heat denaturation, manifested in a complete loss of enzymatic activity, did not disturb the interaction with the antibody.

### b) Effects of Modification of Amino Acid Side Chains

One of the most widely used approaches to the understanding of structure-function relationships of proteins involves the chemical modification of specific amino acid residues and the study of the resulting effect on biological activity. Indeed, the chemical investigations of LANDSTEINER (1945) on the specificity of serological reactions are based on the introduction of antigenic determinants by chemical modification of the protein antigen. Numerous methods have been used for modification, and several enzymes have been investigated in an effort to identify those residues which partake in activity. Immunological studies of chemically modified enzymes are useful for the correlation of catalytic activity, which is confined within only one site on the enzyme, and the antigenic properties to which numerous determinants are contributing.

Experiments with bovine pancreatic ribonuclease have demonstrated that three different amino acid residues, histidine 12, histidine 119 and lysine 41, are essential for the maintenance of catalytic activity. Derivatives in which any one of these residues is modified are enzymatically inactive (STARK et al., 1961; HIRS et al., 1961). At the same time BROWN (1963) has shown that in the immunological reaction with anti-RNase these inactive derivatives behaved identically to native RNase, implying that these residues do not play a part in the immunogenic properties of the molecule. On the other hand, poly-DL-alanyl ribonuclease retains catalytic activity — with suitable substrates it is even more active than native RNase (WELLNER et al., 1963)—but its capacity to interact with antibodies to the native enzyme is considerably lower than that of the native enzyme (BROWN et al., 1963). The attachment of poly-DL-alanyl peptide chains to the ε-amino groups was found to affect trypsin in a similar manner (EPSTEIN et al., 1962; ARNON and NEURATH, 1970), the catalytic activity of the enzyme being only slightly reduced, whereas the capacity to react with the antibodies decreased drastically. A similar effect, albeit less pronounced, was also observed when the amino groups of trypsin were modified by guanidination (ROBINSON and WALSH, 1968;

ARNON and NEURATH, 1970). On the other hand, modification of the carboxylic groups of trypsin by attachment of either glycine ester or poly-DL-alanine chains, reduced the catalytic activity while not affecting at all the interaction with the antibodies (ARNON and NEURATH, 1970).

The involvement of the tyrosine residues in both the catalytic and antigenic activities was studied in the cases of trypsin, lysozyme and nuclease. The tyrosines were modified by nitration. In the case of trypsin, nitration of up to six of the ten tyrosine residues had virtually no effect on the catalytic activity (KENNER et al., 1968) but did diminish the immunological interaction (ARNON and NEURATH, 1970). A similar observation was reported for lysozyme: modification of three out of the six tyrosine residues of the molecule brought about a limited decrease in the immunological capacity, but did not impair the enzymatic activity (BONAVIDA, 1968). Upon nitration of staphylococcal nuclease, an interesting phenomenon was observed: two mononitro derivatives of nuclease — one of which (nitrated at tyrosine residue 85) is enzymatically inactive and the other (nitrated at tyrosine 115) is an active derivative (CUATRECASAS et al., 1968b) — were studied immunologically (FUCHS et al., 1969). Although both these derivatives reacted readily with the antibodies, $Ca^{++}$ and substrate analogues, such as deoxythymidine diphosphate, were capable of interfering with the immunospecific precipitation only of the active derivative (an effect observed with the native enzyme) but had no effect on the enzymatically inert derivative. These findings indicate that neither of these two tyrosine residues is involved in the interaction with the antibodies, but that the conformational "rigidification" accompanying the binding of $Ca^{++}$ and substrate analogues either to the native enzyme or to the active nitrated derivative, changes the availability of some antigenic determinants in their proximity.

Other studies on chemically modified derivatives of nuclease were concerned with the modification of the amino groups (OMENN et al., 1970b). The amino groups were modified by either acetylation or trifluoroacetylation, modifications which induce drastic changes of the surface charge of the protein. The results indicated that an increasing extent of substitution was accompanied by a parallel decrease in the enzymatic and antigenic activities. On the other hand, performic acid-oxidized nuclease, which according to spectral measurements appeared to be denatured, and retained about 8% of its enzymic activity, cross-reacted *fully* with antibodies to the native enzyme. This is in contrast with the results obtained with other performic acid-oxidized enzymes such as ribonuclease (BROWN et al., 1959). However, it should be borne in mind that staphylococcal nuclease lacks disulfide bonds so that the effect of oxidation with performic acid consists of modification of the tryptophan and methionine residues, rather than cleavage of disulfide bonds and unfolding of the peptide chain.

Derivatives of lysozyme, modified at different side chains of specific amino acid residues, have also been studied immunologically. For example, a derivative in which the six tryptophan residues had been modified with 2-nitrophenylsulphenyl chloride (HABEEB and ATASSI, 1969) lost its enzymatic activity completely, concurrently with a drastic decrease, by 82%, of its ability to react with antibodies to the native enzyme. The great reduction in the immunological activity of this derivative might reflect true involvement of one or more of the tryptophan residues in antigenic reactive regions, or might simply be the result of conformational changes shown to take place upon modification. STROSBERG and KANAREK (1969) observed that modification of five tryptophan residues by formylation caused a less drastic change. BONAVIDA (1968) found that when only tryptophan residue 62 of lysozyme was modified by reaction with N-bromosuccinimide (HAYASHI et al., 1965), the derivative, although enzymatically inactive, was fully reactived with anti-lysozyme antibodies, and so was another derivative, retaining 60% of the catalytic activity, in which either tryptophan residue 62 or 63 was modified. On the other hand, in accordance with the results mentioned previously concerning trypsin, guanidination or succinylation of all the amino groups in lysozyme (HABEEB, 1967), or their modification by acetylation or carbamylation (STROSBERG and KANAREK, 1968), resulted in a considerable decrease in the ability to react with antibodies to native lysozyme.

In conclusion, in most cases there is no correlation between the effect on catalytic and immunological activities induced by modification of specific side chains. It should be emphasized that this approach has been useful in certifying that the changes in antigenic modification are not necessarily due to changes in the conformation of the molecule, but may rather be due to limited local disturbances in the enzyme structure.

## 6. Activation of Proenzymes

### a) Enzyme-Proenzyme Correlation

As already mentioned, many enzymes, mainly animal proteases, among them pepsin, chymotrypsin, trypsin, carboxypeptidase, thrombin and plasmin, exist within the tissues in the form of inactive precursors called proenzymes or zymogens. Their activation, or conversion to the active form, involves the splitting of some peptide bonds, a process which in many cases is catalyzed by special enzymes — the kinases. This mechanism provides means of control over the amount of active enzyme present in the tissues at any given time. The activation process sometimes involves the removal of only a small peptide, such as in the conversion of trypsinogen to trypsin, for example, in which a hexapeptide is released, whereas in other cases the remaining unit itself is only a fraction of the molecule — in the activation of procarboxypeptidase A, two-thirds of the molecule are removed. The immunological studies of enzymes and their precursors and of the relationship between them may therefore both serve to help in their characterization and to shed light on the conformational changes involved in the activation process. The immunological correlation between several enzymes and their precursors has now been investigated and will be described in the following.

The first system to be discussed is that of pepsin and pepsinogen. In several respects this is a unique system due to the extreme difference in the nature of the two proteins involved. Native pepsin, containing 71 dicarboxylic amino acid residues and only four basic residues (BLUMENFELD and PERLMANN, 1959), is stabile and active only at very low pH, and is irreversibly denatured at pH values more alkaline than 6. Pepsinogen, on the other hand, is stable in the pH range of 6—9, and at pH values below 5 is converted to pepsin, presumably by an autocatalytic reaction, releasing several peptides. Antibodies can be prepared against both pepsinogen and pepsin. The anti-pepsin antibodies inhibit pepsin activity only to a slight extent (NORTHROP, 1930), but this may be due to the fact that in the immunization process the native enzyme is probably rapidly inactivated at the physiological pH. The serological relationship between pepsin and pepsinogen was studied in several laboratories, leading to different conclusions: SEASTONE and HERRIOTT (1937), using both pepsin and pepsinogen as immunizing agents, observed merely a weak cross-reaction between the two proteins. Similar results were obtained by ARNON and PERLMANN (1963 a, b). LOBACHEVSKAYA (1956), on the other hand, was able to demonstrate cross-precipitation between anti-pepsinogen and pepsin, but only when the latter was prepared from pepsinogen and, therefore, FREEDBERG et al. (1962) suggested that the cross-reacting substance was probably not pepsin itself, but rather the pepsin-inhibitor complex formed in the course of activation. SCHLAMOWITZ et al. (1963) agreed with these findings and suggested that the role played by the inhibitor is to prevent denaturation of pepsin at the high pH prevalent in the serological reaction, since they found that the extent of cross-reactivity between anti-pepsinogen and pepsin can be increased if the reaction is carried out at low pH, when pepsin is maintained in its native form. In agreement with this observation, GERSTEIN et al. (1964) found that the extent of cross-reaction between pepsinogen and antipepsin is drastically increased upon heat denaturation of the zymogen, which presumably unmasks the pepsin moiety in the proenzyme (Fig. 7). From these accumulated data it must be concluded that in this case, due to the major difference in their character and conformation, the enzyme and the zymogen in their native forms show only a low extent of cross-reactivity, and that cross-reactions are observed only upon introduction of appropriate conformational changes.

Different behaviour was observed in the activation of chymotrypsinogen to chymotrypsin, which is known to involve the release of a relatively short peptide (NEURATH, 1957; DESNUELLE, 1960). RICKLY and CAMPBELL (1963) prepared precipitating antibodies against both chymotrypsinogen and $\alpha$-chymotrypsin. On the basis of cross-reaction between the two proteins, they concluded that the two types of antibodies were distinct species; immunization with the proenzyme alone, however, elicited the production of both types of antibodies, and they suggested that this might be due to *in vivo* activation. BARRETT and THOMPSON (1965) observed that antibodies to chymotrypsinogen yielded cross-reaction of identity with a series of $\alpha$, $\beta$, $\gamma$ and $\delta$-chymotrypsins which had been prepared by tryptic activation of chymotrypsinogen under different conditions. Since it is improbable that all these variations of the enzyme are formed simultaneously in the *in vivo* activation of the proenzyme, the authors suggested that the identical cross-reaction is due to the close structural relationship of these enzymes, which places them beyond the limit of recognition by a single antiserum, and that the antibodies to the proenzyme indeed cross-react with the active forms of the enzyme.

The activation of trypsinogen to trypsin, a process which involves the mere release of a hexapeptide from the amino terminus of the molecule (NEURATH, 1957) does not lead to any alteration in the immunological properties of the

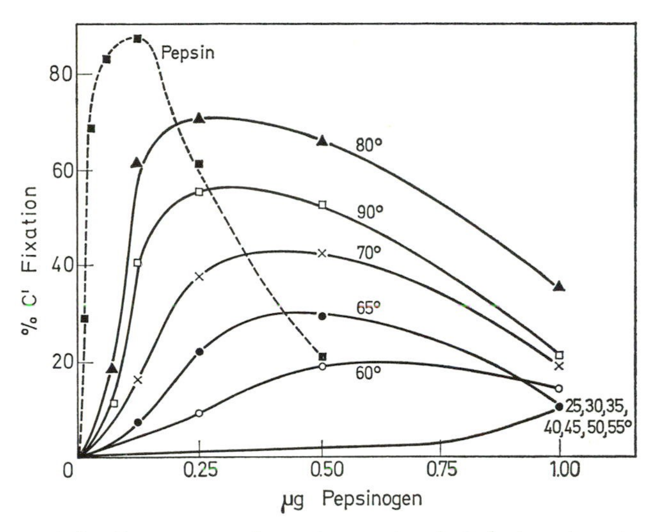


Fig. 7. Complement fixation curves of pepsinogen incubated at various temperatures and assayed with anti-pepsin. Source: VAN VUNAKIS and LEVINE (1963)

molecule. Thus anti-trypsin was found not to distinguish between trypsin and trypsinogen in gel diffusion (BARRETT et al., 1967), and when compared quantitatively, bovine trypsin and bovine trypsinogen yielded identical quantitative precipitin reactions with antisera elicited by either protein (ARNON and NEURATH, 1970). It must be concluded, therefore, that in this case the removal of the activation peptide from the zymogen does not lead to depletion of antigenic determinants, nor does it bring about unmasking of such determinants.

Similarly, during the activation of human plasminogen to plasmin the enzyme-proenzyme pair were found to react identically in gel diffusion with antibody to either proenzyme or enzyme, and antibodies specific towards the plasmin or towards the plasminogen were found to neutralize plasmin to a similar extent (ROBBINS and SUMMARIA, 1966). This is, therefore, another example of a proenzyme-to-enzyme conversion which probably involves only minimal conformational changes resulting from the liberation of a small peptide

(ROBBINS et al., 1965), and consequently does not influence the antigenic properties of the molecule.

The last system to be described here is carboxypeptidase A. In this case the activation of the proenzyme entails a considerable change in the original material, since the molecule of the active enzyme constitutes only about one-third of the procarboxypeptidase A molecule (YAMASAKY et al., 1963). Nevertheless, a high degree of cross-precipitation between the two components and complete cross-reactivity by passive cutaneous anaphylaxis have been observed (BARRETT, 1965a; BEATY, 1966). Moreover, the anti-proenzyme was shown to cause marked inhibition of the enzymatic activity of carboxypeptidase A. On the other hand, antibodies to carboxypeptidase A were found not to react directly with procarboxypeptidase A in the complement fixation technique (LEHRER and VAN VUNAKIS, 1965) and the similarity in their antigenic structure could be demonstrated only by the capacity of the proenzyme to inhibit the enzyme — anti-enzyme reaction. Consequently, it can be inferred that many of the antigenic determinants of procarboxypeptidase A are also to be found on the surface of carboxypeptidase A, and that despite the drastic reduction in molecular weight a structural similarity between the two substances is maintained. It is of interest to note that these results are in accord with the findings obtained in biochemical studies (BROWN et al., 1963).

This behaviour of the procarboxypeptidase A system is in sharp contrast to that of procarboxypeptidase B (of porcine origin). In the latter case the anti-proenzyme serum was found to be extremely specific, and to react only with the homologous antigen and not with the active enzyme it generates (BARRETT, 1965b).

### b) Effects of Antibodies on Activation of Proenzymes

In several cases antibodies to proenzymes have been reported to prevent the activation of the proenzyme and the release of the active enzyme from the complex. Thus, the activation of pepsinogen or modified pepsinogen was inhibited by the interaction with the corresponding antisera (VAN VUNAKIS et al., 1963; ARNON and PERLMANN, 1963a), and procarboxypeptidase did not yield free carboxypeptidase when the activation was attempted in the presence of the antibodies (BARRETT, 1965a). The activation of chymotrypsinogen was partially inhibited by the antibodies to the proenzyme (BARRETT and THOMPSON, 1965) and so was the conversion of plasminogen to plasmin (ROBBINS and SUMMARIA, 1966). BARRETT and EPPERSON (1967) observed that if the activation procedure is carried out on the proenzyme-antibody complex of both trypsinogen and chymotrypsinogen, and is followed by exposure to conditions which dissociate the complex with the antibody, active enzyme is released. These findings indicate that activation of both trypsinogen and chymotrypsinogen does take place even in the presence of their specific antibodies, but the enzymes are not released unless the antibody is dissociated.

In conclusion, the immunological correlation between proenzymes and the enzymes they generate depends on the nature of the activation process. When extensive differences exist in the nature of the two species, they will drastically differ in their antigenic activity as well, but when the activation process involves more limited changes, the concomitant effects on the immunological activity will be minor. In these cases the antibodies will not intervene with the activation process.

### 7. Enzymes and Apoenzymes

Many enzymes are known to require the interaction with a prosthetic group to be able to exert their catalytic activity. These prosthetic groups may consist of either small organic molecules or metal ions, and can be removed from the complex with the enzyme to form the apoenzyme, which is enzymatically inactive. The conversion of the enzyme to the apoenzyme and the reconstitution of the holoenzyme upon introduction of the prosthetic group have been studied in various cases in view of the conformational changes involved in the process. In several cases the elucidation of this aspect of enzyme chemistry was also attempted by an immunochemical approach, yielding valuable information concerning the role which the prosthetic groups play in maintaining the structure of the native enzyme. A few examples will illustrate this point.

Rabbit muscle glycogen phosphorylase b, which consists of a dimer of two identical subunits, each containing pyridoxal phosphate as the prosthetic group, elicits the production of antibodies in roosters and goats (YUNIS and KREBS, 1962). These antibodies react only to a slight extent ($< 5\%$) with apophosphorylase b, which readily dissociates into monomers. If the apoenzyme is prepared using deforming agents (SHALTIEL et al., 1966) no irreversible changes occur in the protein, and after reconstitution with pyridoxal phosphate the holoenzyme regains not only the full catalytic activity but also 90—110% of its precipitability with anti-phosphorylase b antibodies (SHALTIEL, 1968). The apoenzyme as such was shown to be immunologically indistinguishable from the holoenzyme if the reaction was carried out in 1.5 M NaCl, conditions in which the apoenzyme aggregates (HEDRICK et al., 1966). This could imply that the apoenzyme fails to precipitate in the dissociated state since it behaves as a monovalent antigen. However, several compounds which also cause aggregation of the apoenzyme (e.g., pyridoxal or adenylic acid at $1 \times 10^{-3}$M) fail to promote the cross-reaction. It was suggested, therefore (SHALTIEL, 1968), that the pyridoxal phosphate itself is not part of a major antigenic determinant, but that it is involved in maintaining the structure of the major antigenic determinants of the molecule, which is dependent on the conformation of the protomers.

Another enzyme which involves pyridoxal phosphate as a prosthetic group is glutamic-aspartic transaminase. Here also the holoenzyme is predominantly a dimer of two subunits and the apoenzyme is mainly in the monomeric form (POLYANOVSKY, 1968). However, in this case antiserum specific for the native enzyme reacted with the apoenzyme just as well as with the native enzyme in the gel diffusion test, and in activating the catalytic activity. Similarly, antibodies against the apoenzyme could inhibit the catalytic activity of both apotransaminase and the transaminase to the same extent (PATRAMANI et al., 1969). Moreover, addition of pyridoxal phosphate did not have any effect on the binding of the anti-enzyme to the apoenzyme. It was, therefore, concluded that in this case the intact conformation of the transaminase, as maintained and stabilized by pyridoxal phosphate, is not essential for the enzyme — anti-enzyme binding and that the antibody-producing cells use the same pattern for the production of either anti-transaminase or anti-apotransaminase.

Another enzyme, in which the holoenzyme and the apoenzyme, obtained by the removal of flavine-adenine dinucleotide, reacted in a similar fashion with antibodies produced against both species, is D-amino acid oxidase from hog kidneys. Both in gel diffusion and in the quantitative precipitin reaction identity was demonstrated in the reaction with the two proteins (MIYAKE et al., 1969). A slight difference between the antibodies was observed as far as their capacity to inhibit the catalytic activity of the enzyme was concerned, and by this criterion the anti-holoenzyme was a more efficient inhibitor than the anti-apoenzyme. In view of biochemical studies (MIYAKE et al., 1965; YAGI et al., 1967), which indicated that the size, shape and hydrodynamic properties of the apoenzyme were almost the same as those of the holoenzyme, and in view of the minor differences observed in the immunological studies, it was suggested that only a part of the protein moiety, probably in the vicinity of the active site, was altered by the presence of the prosthetic group.

Another type of prosthetic group includes, as mentioned already, metal atoms which participate in the enzyme structure. The importance of metals in biological systems is attested by the abundancy of metal-protein complexes and metalloenzymes containg various transition elements (VALLEE, 1955). Their presence has made possible the investigation of the functional and structural characteristics of the active sites of the respective enzymes. Immunochemical studies carried out with several zinc-containing enzymes, for example, were helpful in correlating the binding of the metal with the structural conformation of the holoenzyme.

Carbonic anhydrase from mammalian erythrocytes appears in several forms, all of which contain one atom of Zn per protein molecule. The removal of the zinc results in a metal-free, inactive enzyme derivative which, however, has identical immunological reactivity as the native enzyme. Similarly, identical immunological properties are also shared by other metal derivatives of the enzyme containg Cu or Co. The structures of the native enzyme and the apoenzyme, therefore, appear to be quite similar (WISTRAND and RAO, 1968).

Another zinc metalloenzyme very extensively studied during recent years is carboxypeptidase A. The investigation of this enzyme was somewhat more complicated due to the fact that the nature of the enzyme depends on the method in which it was generated from procarboxypeptidase. Four different forms — $\alpha$, $\beta$, $\gamma$ and $\delta$ — can be obtained which differ from each other in the number of amino acid residues in the N-terminal portion, and in some physical properties and probably conformation as well (NEURATH et al., 1968). Two of these forms, the $\gamma$ form (referred to as CPA-Anson) and the $\delta$-form (CPA-Allan) are indistinguishable in their amino acid sequence but still differ from each other, probably in conformation. For example, although the zinc atom is catalytically indispensable in both forms, the CPA-Anson cannot be as readily activated upon addition of zinc to the apoenzyme (VALLEE et al., 1960). The immunochemical studies were carried out with antibodies prepared against CPA-Anson (LEHRER and VAN VUNAKIS, 1965). These antibodies reacted readily with the native form of both CPA-Anson and CPA-Allan. A drastic difference in the immunological interaction was observed, however, between the apoenzymes of the two forms. The metal-free apoenzyme of CPA-Allan, although enzymatically inactive, retained its full capacity to react with the antibodies. Upon readdition of zinc the enzymatic activity returned, while the immunological activity remained unchanged. This is in agreement with findings that the apoenzyme is quite similar to the native enzyme (LUDWIG et al., 1963; NEURATH, 1960). In contrast, CPA-Anson, upon removal of the zinc, lost both its enzymatic activity and its ability to fix complement, and readdition of the zinc only partially restored these two activities. Hence it was implied that in the case of CPA-Anson the apoenzyme is different in its conformation from the native enzyme, and that only part of the molecules can be reconstituted to form the holoenzyme.

O'BRIEN and KÄGI (1968) studied carboxypeptidase A from another aspect. They took advantage of the involvement of zinc in carboxypeptidase in an attempt to explore the participation of the active site or its immediate vicinity as antigenic determinants. Assuming that the zinc atom is at the active site, they followed the effect of antibodies on the rate of isotope exchange of this Zn. Indeed, antibodies brought about retardation of the zinc exchange, an effect which paralleled their inhibiting capacity but not their precipitability.

In conclusion, the prosthetic group in all the above examples, whether it be an organic molecule or a metal atom, does not appear as such to play any role as an antigenic determinant. It will affect the antigenic reactivity of the enzyme only in those cases when it is essential for maintaining the native structure of the molecule.

## 8. Multiple Forms of Enzymes

### a) Structural Relationships among Isozymes

It is now well recognized that a large number of enzymes exist in multiple forms, not only in tissues but also in the crystalline state. The differences between the various forms of the same enzyme may be due to different combinations of the same number of specific polypeptide subunits, as in the case of lactic dehydrogenase (CAHN et al., 1962); or they may be attributable to simple amino acid replacements, as in the case of bovine carboxypeptidase (NEURATH et al., 1968). Assuming that such differences stem from the genes controlling the synthesis of the enzymes, one might add mutant enzymes to the general family of isoenzymes, and thus the investigation of multiple forms of enzymes may also be tackled by genetic studies.

The recognition of the existence of isoenzymes originally was due to the finding of multiple bands of enzymatic activity in electrophoresis. The existence of immunological differences between the multiple forms provides a much more sensitive method for their detection and study. Indeed, numerous immunological studies on isoenzymes have appeared during the last several years. Only a few examples will be given in the following, to demonstrate the parallelism between the existence of differences in antigenic determinants and the multiplicity of enzyme forms.

Lactic dehydrogenase is probably the most extensively studied enzyme. As isolated from various species, it has been shown to exist as five isoenzymes of differing subunit composition. All five isozymes are enzymatically active, with similar specificities, and they stem from the random tetrameric association of two different subunits, M and H, each encoded in a separate gene (Holmes and Markert (1969). The two isozymes LDH-I ($H_4$) and LDH-V ($M_4$), each consisting of a single type of subunit only, did not show any immunological cross-reaction (Cahn et al., 1962; Lindsay, 1963), but, as shown in Table 3,

Table 3. *Precipitation of various forms of chicken LDH with a limiting amount of antibody*[a]

| LDH form | Percent precipitation | |
|---|---|---|
| | Anti-M | Anti-H |
| $M_4$ | 100 | 0 |
| $M_3H_1$ | 65 | 17 |
| $M_2H_2$ | 32 | 58 |
| $M_1H_3$ | 8 | 95 |
| $H_4$ | 0 | 100 |

[a] The level of each antibody used was that amount which was in slight excess to completely precipitate the pure homologous form. In all cases the level of antibody, as well as of antigen, was kept constant.

Source: Kaplan and White (1963).

antisera to both forms cross-precipitated with the different hybrid isozymes (Kaplan and White, 1963; Markert and Appella, 1963; Rajewsky et al., 1964) and the extent of this cross-reaction was proportional to the contents of the respective subunit in the hybrid. Cross-reaction was also observed between LDH from different species, and it indicated that in contrast with the two different subunits, subunits of the same type originating from a variety of species do show immunological relationship (Wilson et al., 1964). This was very clearly demonstrated by the capacity of rabbit antiserum against pig LDH-I to precipitate rabbit LDH-I, namely the autologous antigen (Rajewsky, 1966). The two enzymes must, therefore, differ slightly from each other to an extent which is sufficient for rendering the pig's enzyme immunogenic in rabbits, but on the other hand, they must possess a number of determinants of similar antigenic specificity. A limited immunological relationship between LDH-I and LDH-V could also be shown, but these similar surface antigenic determinants were revealed only following acetylation of both subunits (RAJEWSKY and MULLER, 1967).

Glutamic dehydrogenase was also found to exist in multiple forms, which constitute a system in rapid monomer-polymer equilibrium. Immunochemical studies (TALAL et al., 1964) have shown that the enzyme is composed of three immunologically distinct forms, which differ from one another in both their antigenic and catalytic properties — one form demonstrates primarily glutamic dehydrogenase activity, the second catalyses predominantly the alanine dehydrogenase reaction, whereas the third form has both enzyme activities. Subsequent studies (TALAL and TOMKINS, 1964) have shown that the antigenic differences are associated with conformational changes between the different forms, since regulator molecules such as ADP or diethylstilbestrol induced changes in the immunological reactivity. The three isozymes, therefore, are associated in this case with different conformational states of glutamic dehydrogenase.

Horse liver alcohol dehydrogenase represents a third type of situation in effect in multiple forms of enzymes. This enzyme also exists as five electrophoretically different subunits, one of which (the major component) possesses only alcohol dehydrogenase activity whereas a different isozyme is additionally active in the catalysis of the interconversion of some keto steroids to the hydroxy compounds. Immunological studies have shown (PIETRUSZKO and RINGOLD, 1969; PIETRUSZKO et al., 1969) that the last mentioned form is a hybrid, whereas the first isozyme consists of two identical units. The two subunits showed a close immunological relationship — since the two isozymes precipitated in a similar fashion with the two antibodies. This is therefore an opposite case to that of lactic dehydrogenase — the two types of subunits are immunologically related but differ in their enzymic properties.

Carbonic anhydrase (HCA) represents a still different situation. This enzyme, as isolated from human red blood cells, contains two isozymes, a highly active HCA C, and a less active form HCA B (NYMAN, 1961). These isozymes were shown to differ also in some of their physical and chemical properties, and indeed were found to differ in their immunological behaviour (WISTRAND and RAO, 1968). The difference in immunological reactivity manifested itself in lack of both cross-precipitation with the antibodies, and cross-inhibition of catalytic activity by the antibodies. On the other hand, antibodies to these two different isozymes were capable of reacting with the two forms of the enzyme isolated from other species (TASHIAN et al., 1965). A comparative study carried out with carbonic anhydrases from many species has indicated that the two forms of the enzyme originated by gene duplication, and that the HCA C represents the original form.

Another instance in which two forms of an enzyme appeared to be antigenically different is the case of myosin, which possesses adenosine triphosphatase activity. Actomyosins are present in two types of fibers, the rapidly contracting white fibers and the red fibers. The two types of enzyme differ in their enzymatic properties — the red fiber enzyme having a lower catalytic activity. Recently it was demonstrated (GROSCHEL-STEWART and DONIACH, 1969) that they also differ in their antigenic properties and show no cross-reaction, and thus can be differentiated by the reaction with their respective antibodies.

It appears, therefore, that in most cases isozymes, although probably originating from a common ancestral gene, developed into species which are immunologically distinct, namely, the structure of their antigenic determinants is different.

### b) Effects of Genetic Alteration on Enzyme Structure

Antibodies to enzymes have been used to detect mutationally altered enzymes which are devoid of, or possess aberrant, enzymatic activity. They have provided an excellent means for characterizing the antigenic and immunogenic properties of these abnormal proteins and for relating these to other properties of the molecule. In several instances the characteristic properties of the mutant proteins have been related to mutational events occurring within specific regions of a structural gene. These points will be exemplified here by studies with three systems — tryptophan synthetase, $\beta$-galactosidase and alkaline phosphatase.

Tryptophan synthetase is an enzyme consisting of two functionally dependent subunits — $\alpha$ and $\beta_2$. Certain *E. coli* mutants have been found which are devoid of normal enzymic activity, but continue to form molecules which are recognized by their antigenic similarity to the normal enzyme. The antigen involved was designated CRM (cross-reacting material), and appears to be characteristic of particular mutants, which are antigenically distinguishable from each other (SUSKIND et al., 1963). Antibodies can be prepared to each of the two subunits, and used for the detection of alterations in various mutants (YANOFSKY, 1963). The antisera specific for the wild-type $\alpha$-subunit were used in order to measure the effects of single and double amino acid substitutions on the antigenic structure of this subunit (MURPHY and MILLS, 1968). These authors found that about half of all the mutants they studied which exhibited reduced binding affinities to the $\beta_2$ subunit, reacted with the antiserum exactly as the wild-type, thus indicating only limited differences, if any, between the tertiary structures of the wild-type and these mutant proteins.

Immunological studies were also applied successfully in the investigation of *E. coli* $\beta$-galactosidase. This enzyme is a very good immunogen, giving rise to precipitating antibodies which do not inhibit its catalytic activity (COHN and TORRIANI, 1952). As in the case of tryptophan synthetase, many mutants of *E. coli* are deficient of enzyme activity, but produce material which cross-reacts with antibody to the wild-type (PERRIN, 1963). FOWLER and ZABIN (1968) studied the distribution of antigenic determinants by analyzing the immunological activity of a series of nonsense, missense and deletion mutants which map throughout the β-galactosidase structural gene. Their results suggest the presence of three classes of antigenic sites — those which reside in the incomplete chains themselves, those present in mutants mapping either near the center of the gene or in its terminal part, and those due to polymeric protein, thus implying the contribution of specific conformation to the immunological activity. ROTMAN and CELADA (1968) have isolated a cross-reacting protein from a lac(—) *E. coli* mutant which was activated 550-fold by antiserum to the wild-type enzyme. The molecular weight of this purified protein was found to be similar to that of native β-galactosidase, implying that, similarly to the active enzyme, it also consists of a tetramer. The activation by the antibodies may thus be visualized to occur *via* a conformational change proceeding as a result of the contact with antibodies to the native enzyme. In later studies (MESSER and MELCHERS, 1969) eleven lac(—) mutants of *E. coli* were isolated, producing β-galactosidase mutant proteins that can be activated by antiserum to the wild-type enzyme. In these cases, the mechanism of activation is probably also related to induced conformational changes; however, preliminary mapping of these mutants on the basis of their position on the β-galactosidase gene shows that the mutants can be classified in three distinct groups. Two of these groups are activated by different antibody populations, implying that two different antigenic sites on the β-galactosidase are involved in the binding of the antibodies, and each group of mutants is correlated to one antigenic site. The wild-type enzyme gives rise to antibody populations which will recognize different sites on various mutants. Characterization of such mutant enzymes and location of the mutational changes in the polypeptide chain may lead to the definition of specific antigenic determinants on the native enzyme.

Another example in which studies with mutant enzymes revealed information about antigenic determinants is the enzyme alkaline phosphatase (COCKS and WILSON, 1969). In this study the authors investigated the interaction of several enzymatically active mutant enzymes, of single amino acid substitutions, with antiserum to the wild-type enzyme. Although the structural changes in these enzymes were small, most of the mutants tested could be distinguished from the wild-type enzyme in their immunological reactivity. The largest difference observed was in the case of the mutant in which lysine was replaced by glutamic acid, implying that the charge might be of importance in determining the antigenic specificity of this enzyme.

It thus appears that enzymes evolved by genetic mutations show differences in their immunological properties as well. In some cases even a single point mutation, yielding differences in one amino acid, will still lead to a concomitant divergence in the antigenic structure of the enzyme.

## 9. Allosteric Enzymes

The last topic to be mentioned, in which immunological studies have revealed information concerning antigenic determinants, involves the allosteric enzymes. It will be represented here by two systems. The first and probably best understood of the allosteric enzymes, is aspartate transcarbamylase (ATCase) of *E. coli*. This enzyme is composed of two different types of protein subunits, the catalytic subunit and the regulatory subunit, each with a specific function. Specific association of two catalytic and four regulatory subunits leads to the formation of the native enzyme (GERHART and SCHACHMAN, 1965, 1968; CHANGEUX and GERHART, 1967). Preparation of antisera to the intact aspartate transcarbamylase and against the two subunits, and their serological analysis has indicated that the two subunits are immunologically distinct moieties (BETHELL et al., 1968). These antibodies were, therefore, used for the study of conformational changes involved in the dissociation-association of this enzyme. It was found that antisera to the intact enzyme reacted more effectively with native aspartate transcarbamylase than with either or both of its isolated subunits. This effect was attributed to the more effective lattice formation with the multichain protein containing all the antigenic determinants on one molecule. On the other hand, antisera to the catalytic subunit showed greater activity with this subunit than with the intact enzyme, and the anti-regulatory subunit also distinguished the free subunit from the bound form in the intact enzyme, implying that the dissociation resulted in unmasking of some dominant antigenic determinants. Dissociation of ATCase with *p*-hydroxy-mercuribenzoate resulted in a decrease in the activity with anti-ATCase and concomitant increase in the reactivity with the anti-catalytic subunit. The two types of antisera could, therefore, serve to detect dissociation and conformational changes in this system (VON FELLENBERG et al., 1968): it has been demonstrated (GERHART and SCHACHMAN 1968; CHANGEUX and RUBIN, 1968) that conformational changes in ATCase result from interactions with ligands. According to the prediction by MONOD et al. (1965) substrates favour a loose, and feedback inhibitors a tight quaternary structure of an allosteric enzyme. Immunological investigation corroborated this prediction (VON FELLENBERG et al., 1968). In these studies it was observed that several ligands, as well as low substrate concentration, increased the rate of dissociation of ATCase by $\beta$-hydroxymercuribenzoate, whereas high concentrations of the substrate decreased the dissociation; both results were in agreement with the expected conformational changes. Moreover, using the specific antisera to the intact enzyme and to the subunits, these authors demonstrated that the effect of a particular ligand was on the conformation of only that subunit with which the ligand interacts. Thus, only the anti-regulatory immune system was affected by the presence of cytidine triphosphate, whereas carbamyl phosphate and succinate influenced the serological activity of the catalytic subunit.

Another enzyme, in the structure of which an allosteric effect plays a role, is alkaline phosphatase from *E. coli*. This enzyme in its catalytically active form is composed of two identical subunits which can be reversibly dissociated (SCHLESINGER and BARRETT, 1965). Antibodies prepared against the denatured form of the enzyme were not reactive with the active alkaline phosphatase but were directed against the subunit, which by itself is devoid of catalytic activity, but is reassociable to yield active protein (SCHLESINGER, 1967). Since the dimerization process, leading to the formation of the native enzyme, involves both the inclusion of a zinc atom and a conformational change in the molecule (SCHLESINGER, 1965), it must be assumed, in this case as well, that the antibodies to both native enzyme and subunit are capable of detecting the dissociation and change of conformation. Moreover, the antibodies specific toward the subunit, in contrast to the antibodies to the native enzyme, prevented the subunit from reassociation, indicating that they recognize on the subunit, and react with, the sites of assembly of this enzyme.

It appears, thus, that immunological investigations may shed light on the conformational changes involved in the allosteric effects.

## 10. Concluding Remarks

The increasing number of studies on antibodies to enzymes, carried out in the last years have shed light on several problems and aspects raised by enzyme research. The data summarized in this article constitute an attempt to demonstrate the fruitfulness of such immunological investigations. They provided evidence, for example, concerning homology and phylogeny of enzymes, which corroborated the conclusions arrived at by the elucidation of their primary structure. On the other hand, they yielded cardinal information regarding the structure and spatial conformation of enzymes and their various derivatives, as well as that of complex constructions such as the allosteric enzymes. Likewise, they revealed detailed knowledge about subtle differences in molecular structure involved in the activation of proenzymes or in the interactions of apoenzymes with their respective prosthetic groups to form the active holoenzymes. Furthermore, these studies provided answers to questions regarding the existence of, and interrelationship among, multiple forms of enzymes, either those found naturally or those induced by genetical mutations.

The field in which the use of anti-enzyme antibodies have provided the most straightforward information is in the elucidation of the structural features of antigenic determinants of proteins. In this the advantage of enzymes over other proteins resides in the availability of homologous species on the one hand, and in the inherent biological activity of this class of proteins on the other. The correlation of the immunological activity in homologous enzymes, such as cytochromes c from different species, clearly indicated the important role played by hydrophobic amino acid residues in originating immunopotent centers. Likewise, the isolation and analysis of immunologically active fragments from the enzymes trypsin and chymotrypsin, suggest that the presence of a prolyl residue may be of prime importance in endowing the regions in their immediate vicinity with antigenic reactivity. Interestingly, a similar conclusion may be arrived at by inspecting the regions which encompass antigenic determinants of other enzymes, and even their unfolded polypeptide chains. The biological activity of the enzymes, on the other hand, enabled the study of possible correlation between their catalytically active centers and the antigenic structure of the molecule. Although no such relationship could be demonstrated, so that in most cases it may be assumed that the active site itself is not directly involved in an antigenic determinant, antibodies to enzymes are, nevertheless, generally capable of inhibiting the catalytic activity of the enzyme, mostly *via* steric, hindrance. Moreover, among homologous enzymes from different species, cross-inhibition by antibodies is a much more pronounced phenomenon than immunological cross-reaction, implying that during the evolutionary process the structural features responsible for the enzymatic activity were preserved.

Finally, the inhibition of enzymic activity by antibodies served for distinction among different antigenic determinants on the enzyme: the multiplicity of antigenic determinants on the surface of antigens has been accepted as one of the main reasons for the extreme heterogeneity of antibodies, even of the most purified antibody preparations. Selective fractionation of the antibodies into inhibitory and noninhibitory species, as demonstrated for papain and chymopapain, provides proof that the inactivation of the enzyme is a result of its interaction with those antibodies which are specific to regions that are related to the active site. Whereas fractionation of antibodies with different determinant specificities is theoretically possible for every antigen, only with the enzyme antigens has it been feasible to fractionate such species and to distinguish between them on the basis of their inhibitory capacities. Such antibody preparations should be useful in mapping the antigenic structure of enzymes, and in the elucidation of the detailed features of antigenic determinants.

## References

ANDERER, F. A.: Versuche zur Bestimmung der serologische determinanten Gruppen des Tabakmosaikvirus. Z. Naturforsch. **18**b, 1010—1014 (1963).

ARNHEIM, N., PRAGER, E. M., WILSON, A. C.: Immunological prediction of sequence differences among proteins. Chemical comparison of chicken, quail and pheasant lysozymes. J. biol. Chem. **244**, 2085—2094 (1969).

ARNHEIM, N., JR., WILSON, A. C.: Quantitative immunological comparison of bird lysozymes. J. biol. Chem. **242**, 3951—3956 (1967).

ARNON, R.: A selective fractionation of anti-lysozyme antibodies of different determinant specificities. Europ. J. Biochem. **5**, 583—589 (1968).

— MARON, E.: Lack of immunological cross-reaction between Bovine α-lactalbumin and Hen's egg-white lysozyme. J. molec. Biol. in press.

— NEURATH, H.: An immunological approach to the study of evolution of trypsins. Proc. nat. Acad. Sci. (Wash.) **64**, 1323—1328 (1969).

— — Immunochemical studies on bovine trypsin and trypsinogen derivatives. Immunochemistry **7**, 241—250 (1970).

— PERLMANN, G. E.: Relation of immunological and enzymic properties to structural modifications of pepsinogen. J. biol. Chem. **238**, 963—968 (1963a).

— — Antibodies to pepsinogen and pepsinogen modifications. Ann. N. Y. Acad. Sci. **103**, 744—750 (1963b).

— SCHECHTER, B.: Immunological studies on specific antibodies against trypsin. Immunochemistry **3**, 451—461 (1966).

— SELA, M.: Antibodies to a unique region in lysozyme provoked by a synthetic antigen conjugate. Proc. nat. Acad. Sci. (Wash.) **62**, 163—170 (1969).

Arnon, R., Shapira, E.: Antibodies to papain. A selective fractionation according to inhibitory capacity. Biochemistry **6**, 3942—3950 (1967).

— — Comparison between the antigenic structure of mutually related enzymes. A study with papain and chymopapain. Biochemistry **7**, 4196—4200 (1968).

Atassi, M. Z., Habeeb, A. F. S. A., Rydstedt, L.: Lack of immunological cross-reaction between lysozyme and α-lactalbumin and comparison of their conformations. Biochim. biophys. Acta (Amst.) **200**, 184—187 (1970).

— Saplin, B. J.: Immunochemistry of sperm whale myoglobin. I. The specific interaction of some tryptic peptides and of peptides containing all the reactive regions of the antigen. Biochemistry **7**, 688—698 (1968).

Barrett, J. T.: An immunological study of procarboxypeptidase A and its enzyme. Immunology **8**, 129—135 (1965a).

— An antibody to porcine carboxypeptidase B. Int. Arch. Allergy **26**, 158—166 (1965b).

— Epperson, M. S.: The activation of chymotrypsinogen in the presence of antisera. Immunochemistry **4**, 497—499 (1967).

— Nilsson, B., Ghiron, C. A.: A serological investigation of the system: trypsinogen-trypsin. Int. Arch. Allergy **31**, 399—402 (1967).

— Thompson, L. D.: Immunochemical studies with chymotrypsinogen A. Immunology **8**, 136—143 (1965).

Beaty, H. N.: Serological reactions of bovine procarboxypeptidase A and carboxypeptidase A. Biochim. biophys. Acta (Amst.) **124**, 362—373 (1966).

Benjamini, E., Young, J. D., Peterson, W. J., Leung, C. Y., Shimizu, M.: Immunochemical studies on tobacco mosaic virus protein. II. The specific binding of a tryptic peptide of the protein with antibodies to the whole protein. Biochemistry **4**, 2081—2085 (1965).

— — Shimizu, M., Leung, C. Y.: Immunochemical studies on the tobacco mosaic virus protein. I. The immunological relationship of the tryptic peptides of tobacco mosaic virus protein to the whole protein. Biochemistry **3**, 1115—1120 (1964).

Bethell, M. R., Fellenberg, R. von, Jones, M. E., Levine, L.: Immunological studies of aspartate transcarbamylase. I. Characterization of the native enzyme, catalytic and regulatory subunit immune systems. Biochemistry **7**, 4315—4322 (1968).

Blake, C. C. F., Koenig, D. F., Mair, G. A., North, A. C. T., Phillips, D. C., Sarma, V. R.: Structure of hen egg-white lysozyme. A three dimensional fourier synthesis at 2 Å resolution. Nature (Lond.) **206**, 757—761 (1965).

Blumenfeld, O. O., Perlmann, G. E.: The amino acid composition of crystalline pepsin. J. gen. Physiol. **42**, 553—561 (1959).

Bonavida, B.: Molecular basis of the serological specificity of hen egg-white lysozyme. Disseration thesis, University Microfilms (1968).

— Miller, A., Sercarz, E. E.: Structural basis for immune recognition of lysozyme. I. Effect of cyanogen bromide on hen egg-white lysozyme. Biochemistry **8**, 968—979 (1969).

Boyd, W. C.: Introduction to immunological specificity. New York: Interscience 1962.

Branster, M., Cinader, B.: The interaction between bovine ribonuclease and antibody: a study of the mechanism of enzyme inhibition by antibody. J. Immunol. **87**, 18—38 (1961).

Brew, K., Campbell, P. N.: The characterization of the whey proteins of guinea-pig milk. The isolation and properties of α-lactalbumin. Biochem. J. **102**, 258—264 (1967).

— Vanaman, T. C., Hill, R. L.: Comparison of the amino acid sequence of bovine α-lactalbumin and hens egg-white lysozyme. J. biol. Chem. **242**, 3747—3749 (1967).

Brew, K., Vanaman, T. C., Hill, R. L.: The role of α-lactalbumin and the A protein in lactose synthetase: A unique mechanism for the control of a biological reaction. Proc. nat. Acad. Sci. (Wash.) **59**, 491—497 (1968).

Brown, J. R., Greenshields, R. N., Yamasaky, M., Neurath, H.: The sub-unit structure of bovine procarboxypeptidase A-6S. Chemical properties and enzymatic activities of the products of molecular disaggregation. Biochemistry **2**, 867—876 (1963).

Brown, R. K.: Studies on the antigenic structure of ribonuclease. III. Inhibition by peptides of antibody to performic acid-oxidized ribonuclease. J. biol. Chem. **237**, 1162—1167 (1962).

— Immunological studies of bovine ribonuclease derivatives. Ann. N. Y. Acad. Sci. **103**, 754—764 (1963).

— Delany, R., Levine, L., Vunakis, H. van: Studies on the antigenic structure of ribonuclease. I. General role of hydrogen and disulfide bonds. J. biol. Chem. **234**, 2043—2049 (1959).

— Trzpis, M. J., Sela, M., Anfinsen, C. B.: Studies on the antigenic structure of ribonuclease. J. biol. Chem. **238**, 3876—3883 (1963).

Browne, W. J., North, A. C. T., Phillips, D. C., Brew, K., Vanaman, T. C., Hill, R. L.: A possible three-dimensional structure of bovine α-lactalbumin based on that of hen's egg-white lysozyme. J. molec. Biol. **42**, 65—86 (1969).

Cahn, R. D., Kaplan, N. O., Levine, L., Zwilling, E.: The nature and development of lactic dehydrogenases. Science **136**, 962—969 (1962).

Canfield, R. E., Liu, A. K.: The disulfide bonds of egg white lysozyme (muramidase). J. biol. Chem. **240**, 1997—2002 (1965).

Cebra, J. J.: Studies on the combining sites of the protein antigen silk fibroin. III. Inhibition of the silk fibroin — anti-fibroin system by peptides derived from the antigen. J. Immunol. **86**, 205—214 (1961).

Changeux, J.-P., Gerhart, J. C.: Allosteric interactions in aspartate transcarbamylase. In: FEBS Symp. on Regulation of Enzyme Activity and Allosteric Interactions. London and New York: Academic Press 1967.

— Rubin, M. M.: Allosteric interactions in aspartate transcarbamylase. III. Interpretation of experimental data in terms of the model of Monod, Wyman and Changeux. Biochemistry **7**, 553—561 (1968).

Cinader, B.: Antibodies against enzymes. Ann. Rev. Microbiol. **11**, 371—390 (1957).

— Immunochemistry of enzymes. Ann. N. Y. Acad. Sci. **103**, 495—548 (1963).

— Antibodies to enzymes — a discussion of the mechanisms of inhibition and activation. In: Antibodies to biologically active molecules (B. Cinader, ed.). Oxford: Pergamon Press 1967.

— Lafferty, K. J.: Mechanism of enzyme inhibition by antibody. A study of the neutralization of ribonuclease. Immunology **7**, 342—362 (1964).

Cocks, G. T., Wilson, A. C.: Immunological detection of single amino acid substitution in alkaline phosphatase. Science **164**, 188—189 (1969).

Cohn, M., Torriani, A. M.: Immunochemical studies with the β-galactosidase and structurally related proteins of *Escherichia coli*. J. Immunol. **69**, 471—491 (1952).

Crumpton, M. J.: Conformational changes in sperm-whale metmyoglobin due to combination with antibodies to apomyoglobin. Biochem. J. **100**, 223—232 (1966).

— The molecular basis of the serological specificity of proteins, with particular reference to sperm whale myoglobin. In: Antibodies to biologically active molecules (B. Cinader, ed.). Oxford: Pergamon Press 1967.

— Wilkinson, J. M.: The immunological activity of some of the chymotryptic peptides of spermwhale myoglobin. Biochem. J. **94**, 545—556 (1965).

Cuatrecasas, P., Fuchs, S., Anfinsen, C. B.: The tyrosyl residues at the active site of staphylococcal nuclease. J. biol. Chem. **243**, 4787—4798 (1968b).

Cuatrecases, P., Taniuchi, H., Anfinsen, C. B.: The structural basis of the catalytic function of staphylococcal nuclease. Brookhaven Symp. Biol. **21**, 172—198 (1968a).
Desnuelle, P.: Chymotrypsin. In: The enzymes (P. D. Boyer, H. Lardy and K. Myrback, eds.), vol. IV. New York: Academic Press 1960.
Enser, M., Shapiro, S., Horecker, B. L.: Immunological studies of liver, kidney and muscle fructose-1,6-diphosphatases. Arch. Biochem. **129**, 377—383 (1969).
Epstein, C. J., Anfinsen, C. B., Sela, M.: The properties of poly-DL-alanyl trypsin and poly-DL-alanyl chymotrypsin. J. biol. Chem. **237**, 3458—3463 (1962).
Fazekas de St. Groth, S.: Steric inhibition: neutralization of a virus-borne enzyme. Ann. N. Y. Acad. Sci. **103**, 674—687 (1963).
Fellenberg, R. von, Bethell, M. R., Jones, M. E., Levine, L.: Immunological studies of aspartate transcarbamylase. II. Effect of ligands on the conformation of the enzyme. Biochemistry **7**, 4322—4329 (1968).
— Levine, L.: Proximity of the enzyme active center and an antigenic determinant of lysozyme. Immunochemistry **4**, 363—365 (1967).
Fowler, A. V., Zabin, I.: β-Galactosidase: Immunological studies of nonsense, missense and deletion mutants. J. molec. Biol. **33**, 35—47 (1968).
Freedberg, I. M., Lehrer, H., Vunakis, H. van: Immunochemical studies on structural changes associated with conversion of pepsinogen to pepsin. Fed. Proc. **21**, 35 (1962).
Fuchs, S., Cuatrecasas, P., Ontjes, D. A., Anfinsen, C. B..: Correlation between the antigenic and catalytic properties of staphylococcal nuclease. J. biol. Chem. **244**, 943—950 (1969).
Fujio, H., Imanishi, M., Nishioka, K., Amano, T.: Antigenic structures of hen egg-white lysozyme II. Significance of the N- and C-terminal region as an antigenic site. Biken's J. **11**, 207—218 (1968a).
— — — — Proof of independency of two antigenic sites in egg-white lysozyme. Biken's J. **11**, 219—223 (1968b).
— Saiki, Y., Imanishi, M., Shinka, S., Amano, T.: Immunochemical studies on lysozyme. III. Cross-reaction of hen and duck lysozyme and their methyl esters. Biken's J. **5**, 201—226 (1962).
Gerhart, J. C., Schachman, H. K.: Distinct subunits for the regulation and catalytic activity of aspartate transcarbamylase. Biochemistry **4**, 1054—1062 (1965).
— — Allosteric interactions in aspartate transcarbamylase. II. Evidence for different conformational states of the protein in the presence and absence of specific ligands. Biochemistry **7**, 538—552 (1968).
Gerstein, J. F., Levine, L., Vunakis, H. van: Altered antigenicity of pepsinogen and pepsin as an index of conformational change: Effect of urea and reducing agent. Immunochemistry **1**, 3—14 (1964).
Gerwing, J., Thompson, K.: Studies on the antigenic properties of egg-white lysozyme. I. Isolation and characterization of a tryptic peptide from reduced and alkylated lysozyme exhibiting haptenic activity. Biochemistry **7**, 3888—3892 (1968).
Givas, J., Centeno, E. R., Manning, M., Sehon, A. H.: Isolation of antibodies to the C-terminal heptapeptide of myoglobin with a synthetic peptide. Immunochemistry **5**, 314—318 (1967).
Groschel-Stewart, U., Doniach, D.: Immunological evidence for human myosin isoenzyme. Immunology **17**, 991—994 (1969).
Habeeb, A. S. F. A.: Quantitation of conformational changes on chemical modification of proteins: use of succinilated protein as a model. Arch. Biochem. **121**, 652—664 (1967).
— Atassi, M. Z.: Enzymic and immunochemical properties of lysozyme. II. Conformation, immunochemistry and enzymic activity of a derivative modified at tryptophan. Immunochemistry **6**, 555—566 (1969).

Haimovich, J., Hurwitz, E., Novik, N., Sela, M.: Preparation of protein bacteriophage conjugates and their use in detection of anti-protein antibodies. Biochim. biophys. Acta (Amst.) **107**, 115 (1970).

Hayashi, K., Imoto, T., Fanatsu, G., Fanatsu, M.: The position of the active tryptophan residue in lysozyme. J. Biochem. (Tokyo) **58**, 227—235 (1965).

Hedrick, J. L., Shaltiel, S., Fischer, E. H.: On the role of pyridoxal 5'-phosphate in phosphorylase. III. Physicochemical properties and reconstitution of apophosphorylase b. Biochemistry **5**, 2117—2125 (1966).

Hirs, C. H. W., Halmann, M., Kycia, J. H.: The reactivity of certain functional groups in ribonuclease A towards substitution by 1-fluoro-2,4-dinitrobenzene. Inactivation of the enzyme by substitution at the lysine residue in position 41. In: Biological structure and function. Proc. First IUB/IUBS Intern. Symposium, Stockholm (T. W. Goodwin and O. Lindberg, eds.), vol. I, p. 41. New York: Academic Press 1961.

Holmes, R. S., Markert, C. L.: Immunochemical homologies among subunits of trout lactate dehydrogenase isozymes. Proc. nat. Acad. Sci. (Wash.) **64**, 205—210 (1969).

Imanishi, M., Fujio, H., Sakakoshi, M., Amano, T.: Antigenic structures of hen egg-white lysozyme. III. Antibody specificity and neutralizing activity. 2nd Symposium on Immunochemistry, Tokyo, Abstracts p. 45 (1968).

— Miyagawa, N., Fujio, H., Amano, T.: Highly inhibitory antibody fraction against enzymic activity of egg-white lysozyme on a small substrate. Biken's J. **12**, 85—96 (1969).

Isagholian, L. B., Brown, R. K.: Interaction of a peptide with antibody to oxidized ribonuclease. Immunochemistry **7**, 167—174 (1970).

Jollès, J., Niemann, B., Hermann, J., Jollès, P.: Differences between the chemical structures of duck and hen egg-white lysozymes. Europ. J. Biochem. **1**, 344—346 (1967).

Kaminski, M.: The analysis of the antigenic structure of protein molecules. Progr. Allergy **9**, 79—157 (1965).

Kaplan, N. O., White, S.: Immunological characteristics of dehydrogenases. Ann. N. Y. Acad. Sci. **103**, 835—848 (1963).

Kenner, R. A., Walsh, K. A., Neurath, H.: The reaction of tyrosyl residues of bovine trypsin and trypsinogen with tetranitromethane. Biochem. biophys. Res. Commun. **33**, 353—360 (1968).

Landsteiner, K.: The specificity of serological reactions, rev. ed. Cambridge, Massachusetts: Harvard University Press 1945.

Lapresle, C., Darieux, J.: Etude de la dégradation de la serumalbumine humaine par un extrait de rate de lapin. III. Modifications immunologiques de l'albumine en fonction du stade de degradation. Ann. Inst. Pasteur **92**, 62—73 (1957).

Lehrer, H. I., Vunakis, H. van: Immunochemical studies on carboxypeptidase A. Immunochemistry **2**, 255—262 (1965).

Lindsay, D. T.: Isozymic patterns and properties of lactate dehydrogenase from developing tissues of the chicken. J. exp. Zool. **152**, 75—88 (1963).

Lobochevskaya, O. V.: A study of the immunochemical properties of pepsin and pepsinogen. Ukrain. Biokhim. Zhur. **28**, 385 (1956).

Ludwig, M., Paul, I., Pawley, G., Lipscomb, W.: The structure of carboxypeptidase A. I. A two-dimensional superposition function. Proc. nat. Acad. Sci. (Wash.) **50**, 282—285 (1963).

Margoliash, E., Nisonoff, A., Reichlin, M.: Immunological activity of cytochrome *c*. I. Precipitating antibodies to monomeric vertebrate. J. Biol. Chem. **245**, 931—939 (1970).

— Reichlin, M., Nisonoff, A.: The relation of immunological activity and primary structure in cytochrome *c*. In: Conformation of biopolymers (G. N. Ramachandran, ed.). New York: Academic Press 1967.

MARKERT, C. L., APPELLA, E.: Immunochemical properties of lactate dehydrogenase isozymes. Ann. N. Y. Acad. Sci. **103**, 915—929 (1963).
MARON, E., ARNON, R., SELA, M., PERIN, J.-P., JOLLÈS, P.: Immunological comparison of bird and human lysozymes and of their "loop" regions. Biochim. biophys. Acta (Amst.) **214**, 222—224 (1970a).
— SHIOZAWA, C., ARNON, R., SELA, M.: Further studies on a unique antigenic region in lysozyme. Israel J. med. Sci. **6**, 443 (1970b).
MARRACK, J. R.: Enzymes and immunology. In: The enzymes (J. B. Sumner, ed.), vol. 1, pt. 1, p. 343. New York: Academic Press, 1950.
MARSHALL, M., COHEN, P. P.: An immunological study of carbamyl phosphate synthetase. J. biol. Chem. **236**, 718—724 (1961).
McGEACHIN, R. L., REYNOLDS, J. M.: Inhibition of amylases by rooster antisera to hog pancreatic amylase. Biochim. biophys. Acta (Amst.) **39**, 531—532 (1960).
MESSER, W., MELCHERS, F.: The activation of mutant $\beta$-galactosidase by specific antibodies. In: The lactose operon (J. R. BECKWITH and D. ZIPSER, eds.). Cold Spring Harbor Monographs (1969).
MICHAELI, D., PINTO, J. D., BENJAMINI, E.: Immunoenzymology of acetylcholinesterase. II. Effect of antibody on heat-denatured enzyme. Immunochemistry **6**, 371—378 (1969b).
— — — BUREN, F. P. DE: Immunoenzymology of acetylcholinesterase (AChE). I. Substrate specificity and heat stability of AChE and of AChE-anti AChE complex. Immunochemistry **6**, 101—109 (1969a).
MICHAELIDES, M. C., SHERMAN, R., HELMREICH, E.: The interaction of muscle phosphorylase with soluble antibody fragments. J. biol. Chem. **239**, 4171—4181 (1964).
MIYAKE, Y., AKI, K., HASHIMOTO, S., YAMANO, T.: Crystallization and some properties of D-amino acid oxidase apoenzyme. Biochim. biophys. Acta (Amst.) **105**, 86—99 (1965).
— YAMAJI, K., YAMANO, T.: An immunochemical study of D-amino acid oxidase. J. Biochem. (Tokyo) **65**, 531—537 (1969).
MONOD, J., WYMANN, J., CHANGEUX, J.-P.: On the nature of allosteric transitions: A plausible model. J. molec. Biol. **12**, 88—118 (1965).
MURPHY, T. M., MILLS, S. E.: Immunochemical comparisons of mutant and wild-type $\alpha$-subunits of tryptophan synthetase. Arch. Biochem. **127**, 7—16 (1968).
NAJJAR, V. A., FISHER, J.: The mechanism of antibody-antigen reaction. Biochim. biophys. Acta (Amst.) **20**, 158—169 (1956).
NEUMANN, H., STEINBERG, I. Z., BROWN, J. B., GOLDBERGER, R. F., SELA, M.: On the non-essentiality of two specific disulfide bonds in ribonuclease for its biological activity. Europ. J. Biochem. **3**, 171—182 (1967).
NEURATH, H.: The activation of zymogens. Advanc. Protein Chem. **12**, 319—385 (1957).
— Carboxypeptidases A and B. In: The enzymes, vol. 4 (P. D. Boyer, H. Lardy, and K. Myrback, eds.). New York and London: Academic Press 1960.
— BRADHSAW, R. A., ERICSSON, L. H., RABIN, D. R., PETRA, P. H., WALSH, K. A.: Current status of chemical structure of bovine pancreatic carboxypeptidase A. Brookhaven Symp. Biol. **21**, 1—23 (1968).
NG, C. W., GREGORY, K. F.: Antibody to lactate dehydrogenase IV: Enzyme-inhibiting and inhibition-interfering components. Biochim. biophys. Acta (Amst.) **192**, 258—264 (1969).
NISONOFF, A., REICHLIN, M., MARGOLIASH, E.: Immunological activity of cytochrome *c*. II. Localization of a major antigenic determinant of human cytochrome *c*. J. biol. Chem. **245**, 940—946 (1970).
NISSELBAUM, J. S., BODANSKY, O.: Reactions of human tissue lactic dehydrogenases with antisera to human heart and liver lactic dehydrogenases. J. biol. Chem. **236**, 401—404 (1961).

Northrop, J. H.: Crystalline pepsin. I. Isolation and tests of purity. J. gen. Physiol. **13**, 739—766 (1930).

Nyman, P. O.: Purification and properties of carbonic anhydrase from human erythrocytes. Biochim. biophys. Acta (Amst.) **52**, 1—12 (1961).

O'Brien, T. F., Kagi, J. H. R.: The effect of specific antibody on zinc exchange at the active site of carboxypeptidase A. Fed. Proc. **27**, 260 (1968).

Okada, Y., Ikenaka, T., Yagura, T., Yamamura, Y.: Immunological heterogeneity of rabbit antibody fragments against taka-amylase A. J. Biochem. (Tokyo) **54**, 101—102 (1963).

Omenn, G. S., Ontjes, D. A., Anfinsen, C. B.: Immunochemistry of staphylococcal nuclease. II. Inhibition and binding studies with sequence fragments. Biochemistry **9**, 313—321 (1970a).

— — — Immunochemistry of staphyloccal nuclease. I. Physical, enzymatic and immunological studies of chemically modified derivatives. Biochemistry **9**, 304—312 (1970b).

Patramani, I., Katsiri, K., Pistevoi, E., Kologerakos, T., Pavlatos, M., Evanglopoulos, A. E.: Glutamic-aspartic transaminase-antitransaminase interaction. A method for antienzyme purification. Europ. J. Biochem. **11**, 28—36 (1969).

Perrin, D.: Immunological studies with genetically altered $\beta$-galactosidase. Ann. N. Y. Acad. Sci. **103**, 1058—1066 (1963).

Phillips, D. C.: The hen egg-white lysozyme molecule. Proc. nat. Acad. Sci. (Wash.) **57**, 484—495 (1967).

Pietruszko, R., Ringold, H. J.: Antibody studies with the multiple enzymes of horse liver alcohol dehydrogenase. I. Biochem. biophys. Res. Commun. **33**, 497—502 (1969).

— — Kaplan, N. O., Everse, J.: Antibody studies with the multiple enzymes of horse liver alcohol dehydrogenase. II. Biochem. biophys. Res. Commun. **33**, 503—507 (1969).

Pollock, M. R.: An immunological study of the constitutive and the penicillin-induced penicillinases of *Bacillus cereus*, based on specific enzyme neutralization by antibody. J. gen. Microbiol. **14**, 90—108 (1956).

— Stimulating and inhibiting antibodies for bacterial penicillinase. Immunology **7**, 707—723 (1964).

— Fleming, J., Petrie, S.: The effects of specific antibodies on the biological activities of wild-type bacterial penicillinases and their mutationally altered analogues. In: Antibodies to biologically active molecules (B. Cinader, ed.). Oxford: Pergamon Press 1967.

Polyanovsky, O. L.: Pyridoxal catalysis: Enzymes and model systems, p. 155. New York: Interscience 1968.

Porter, R. R.: The isolation and properties of a fragment of bovine serum albumin which retains the ability to combine with rabbit antiserum. Biochem. J. **66**, 677—686 (1957).

Pressman, D., Grossberg, A. L.: The structural basis of antibody specificity. New York: W. A. Benjamin, Inc. 1968.

Rajewsky, K.: Rabbit antibody to pig lactic dehydrogenase reacting with the rabbit's own enzyme. Immunochemistry **3**, 487—489 (1966).

— Avrameas, S., Grabar, P., Pfleiderer, G., Wachsmuth, E. D.: Immunologische Spezifität von Lactatdehydrogenase-Isozymen dreier Säugetier-Organismen. Biochim. biophys. Acta (Amst.) **92**, 248—259 (1964).

— Müller, B.: Similar surface areas on acetylated lactic dehydrogenases. Immunochemistry **4**, 151—156 (1967).

Rickli, E. E., Campbell, D. H.: Antibodies against chymotrypsinogen and $\alpha$-chymotrypsin. Fed. Proc. **22**, 555 (1963).

ROBBINS, K. C., SUMMARIA, L.: An immunochemical study of human plasminogen and plasmin. Immunochemistry **3**, 29—40 (1966).
— — HSIEH, B., LING, C.: Studies on the mechanism of activation of human plasminogen by urokinase. Thrombos. Diathes. haemorrh. (Stuttg.) **13**, 586—587 (1965).
ROBINSON, N. C., WALSH, K. A.: Activation of guianidinated trypsinogen. Fed. Proc. **27**, 292 (1968).
ROTMAN, M. B., CELADA, F.: Antibody-mediated activation of a defective β-D-galactosidase extracted from an *Escherichia coli* mutant. Proc. nat. Acad. Sci. (Wash.) **60**, 660—667 (1968).
SAMUELS, A.: Immunoenzymology-reaction processes, kinetics and the role of conformational alterations. Ann. N. Y. Acad. Sci. **103**, 858—889 (1963).
SANDERS, M. M., WALSH, K. A., ARNON, R.: Immunological cross-reaction between trypsin and chymotrypsin as a guide to structural homology. Biochemistry **9**, 2356—2363 (1970).
SCHLAMOWITZ, M., VARANDANY, P. T., WISSLER, F. C.: Pepsinogen and pepsin: Conformational relations, studied by iodination, immunochemical precipitations and influence of pepsin inhibitor. Biochemistry **2**, 238—246 (1963).
SCHLESINGER, M. J.: The reversible dissociation of the alkaline phosphatase of *Escherichia coli*. II. Properties of the subunit. J. biol. Chem. **240**, 4293—4298 (1965).
— The reversible dissociation of the alkaline phosphatase of *Escherichia coli*. III. Properties of antibodies directed against the subunits. J. biol. Chem. **242**, 1599—1603 (1967).
— BARRETT, K.: The reversible dissociation of the alkaline phosphatase of *Escherichia coli*. I. Formation and reactivation of subunits. J. biol. Chem. **240**, 4284—4292 (1965).
SCIBIENSKI, B., MILLER, A., BONAVIDA, B., SERCARZ, E.: Cross-reaction among related gallinaceous lysozymes. Submitted to Immunochemistry (1970).
SEASTONE, C. V., HERRIOTT, R. M.: Immunological studies on pepsin and pepsinogen. J. gen. Physiol. **20**, 797—806 (1937).
SELA, M.: Immunological studies with synthetic polypeptides. Adv. in Immunol. **5**, 29—129 (1966).
— SCHECHTER, B., SCHECHTER, I., BOREK, F.: Antibodies to sequential and conformational determinants. Cold Spr. Harb. Symp. quant. Biol. **32**, 537—545 (1967).
SHALTIEL, S.: An immunochemical comparison between phosphorylase b and its apoenzyme. Israel J. Chem. **6**, 104p (1968).
— HEDRICK, J. L., FISCHER, E. H.: On the role of pyridoxal 5′-phosphate in phosphorylase. II. Resolution of rabbit muscle phosphorylase b. Biochemistry **5**, 2108—2116 (1966).
SHAPIRA, E., ARNON, R.: The mechanism of inhibition of papain by its specific antibodies. Biochemistry **6**, 3951—3956 (1967).
— — Cleavage of one specific disulfide bond in papain. J. biol. Chem. **244**, 1026—1032 (1969).
SHINKA, S., IMANISHI, M., KUWAHARA, O., FUJIO, H., AMANO, T.: Immunochemical studies on lysozyme. II. On the non-neutralizing antibodies. Biken's J. **5**, 181—200 (1962).
— — MIYAGAWA, N., AMANO, T., INOUYÉ, M., TSUGITA, A.: Chemical studies on antigenic determinants of hen egg-white lysozyme. I. Biken's J. **10**, 89—107 (1967).
SIGLER, P. B., BLOW, D. M., MATTHEWS, B. W., HENDERSON, R.: Structure of α-chymotrypsin. II. A preliminary report including a hypothesis for the activation mechanism. J. molec. Biol. **35**, 143—164 (1968).

Stark, G. R., Stein, W. H., Moore, S.: Relationships between the conformation of ribonuclease and its activity toward iodoacetate. J. biol. Chem. **236**, 436—442 (1961).

Stratton, J., Bonavida, B., Sercarz, E.: Tolerance induction to nonimmunogenic epitopes on hen lysozyme. To be submitted (1970).

Strosberg, A. D., Kanarek, L.: Etude de l'antigenicite du lysozyme. III. Role des residues lysine. Arch. int. Physiol. Biochem. **76**, 949 (1968).

— — Immunochemical studies on hen's egg-white lysozyme. Effect of formylation of tryptophan residues. FEBS Letters **5**, 324—326 (1969).

Suskind, S. R., Wickham, M. L., Carsiotis, M.: Antienzymes in immunogenetic studies. Ann. N. Y. Acad. Sci. **103**, 1106—1127 (1963).

Suzuki, T., Peliochova, H., Cinader, B.: Enzyme activation by antibody. I. Fractionation of Immunsera in search for an enzyme activating antibody. J. Immunol. **103**, 1366—1376 (1969).

Szeinberg, A., Zoreff, E., Golan, R.: Immunochemical gel diffusion study of relationships between erythrocyte catalase of various species. Biochim. biophys. Acta (Amst.) **188**, 287—294 (1969).

Talal, N., Tomkins, G. M.: Antigenic differences associated with conformational changes in glutamate dehydrogenase. Biochim. biophys. Acta **89**, 226—231 (1964).

— — Mushinski, J. F., Yielding, K. L.: Immunochemical evidence for multiple molecular forms of crystalline glutamic dehydrogenase. J. molec. Biol. **8**, 46—53 (1964).

Taniuchi, H., Anfinsen, C. B.: Steps in the formation of active derivatives of staphylococcal nuclease during trypsin digestion. J. biol. Chem. **243**, 4778—4786 (1968).

Tashian, R. E.: Genetic variation and evolution of the carboxylic esterase and carbonic anhydrases of primate erythrocytes. Amer. J. hum. Genet. **17**, 257—272 (1965).

Vallee, B. L.: Zinc and metalloenzymes. Advanc. Prot. Chem. **10**, 317—384 (1955).

— Coombs, T. L., Hoch, F. L.: The "active site" of bovine pancreatic carboxypeptidase A. J. biol. Chem. **235**, PC45—PC47 (1960).

Vunakis, H. van, Lehrer, H. I., Allison, W. S., Levine, L.: Immunochemical studies on the components of the pepsinogen system. J. gen. Physiol. **46**, 589—604 (1963).

— Levine, L.: Structural studies on pepsinogen and pepsin: an immunological approach. Ann. N. Y. Acad. Sci. **103**, 735—743 (1963).

Walsh, K. A., Neurath, H.: Trypsinogen and chymotrypsinogen as homologous proteins. Proc. nat. Acad. Sci. (Wash.) **52**, 884—889 (1964).

Wellner, D., Silman, H. I., Sela, M.: Enzymatic activity of polyalanyl ribonuclease. J. biol. Chem. **238**, 1324—1331 (1963).

Weston, D.: A specific antiserum to lysosomal cathepsin D. Immunology **17**, 421—428 (1969).

Wilson, A. C., Kaplan, N. O., Levine, L., Pesce, A., Reichlin, M., Allison, W. S.: Evolution of lactic dehydrogenases. Fed. Proc. **23**, 1258—1266 (1964).

Wistrand, P. J., Rao, S. N.: Immunological and kinetic properties of carbonic anhydrases from various tissues. Biochim. biophys. Acta (Amst.) **154**, 130—144 (1968).

Yagi, K., Naoi, M., Haradh, M., Okamura, K., Hikada, H., Ozawa, T., Kotaki, A.: Structure and function of D-amino acid oxidase. I. Further purification of hog kidney D-amino acid oxidase and its hydrodynamic and optical rotatory properties J. Biochem. (Tokyo) **61**, 580—597 (1967).

Yamasaki, M., Brown, J. R., Cox, D. J., Greenshield, R. N., Wade, R. N., Neurath, H.: Procarboxypeptidase A-S6. Further studies of its isolation and properties. Biochemistry **2**, 859—866 (1963).

Yanofsky, C.: Antibodies to the two protein components of *Escherichia coli* tryptophan synthetase. Ann. N. Y. Acad. Sci. **103**, 1067—1074 (1963).

Young, J. D., Leung, C. Y.: Immunological studies on lysozyme and carboxymethylated lysozyme. Fed. Proc. **28**, 326 (1969).

Yunis, A. A., Krebs, E. G.: Comparative studies on glycogen phosphorylase. II. Immunological studies on rabbit and human skeletal muscle phosphorylase. J. biol. Chem. **237**, 34—39 (1962).

Zyk, N., Citri, N.: The interaction of penicillinase with penicillins. IV. Comparison of free and antibody-bound enzyme. Biochim. biophys. Acta (Amst.) **99**, 427—441 (1965).

— — The interaction of penicillinase with penicillins. VI. Comparison of free and antibody-bound enzyme. Biochim. biophys. Acta (Amst.) **159**, 317—326 (1968a).

— — The interaction of penicillinase with penicillins. VII. Effect of specific antibodies on conformative response. Biochim. biophys. Acta (Amst.) **159**, 327—339 (1968b).

Max-Planck-Institut für experimentelle Medizin, Abteilung für Chemie, Göttingen

# The Genetic Apparatus of Mitochondria from *Neurospora* and Yeast

Hans Küntzel

With 5 Figures

Contents



## I. Introduction

The function, biogenesis and genetic autonomy of mitochondria has remained a central topic for biologists and biochemists since the early speculations at the end of the 19th century (Altmann, 1890). The discovery of mitochondrial DNA in the late 1950 (Chevremont et al., 1959; Nass and Nass, 1962) initiated a new phase of research which resulted in a logarithmic growth of literature on mitochondrial biogenesis and in a parallel production of review articles (Gibor and Granick, 1964; Wilkie, 1964; Roodyn and Wilkie, 1968; Wagner, 1969; Nass, 1969a; Schatz, 1970).

The aim of this article is to discuss in more details some new findings concerning the molecular genetics of mitochondria from *Neurospora crassa* and yeast. Although mitochondria from vertebrates are better characterized than microbial mitochondria with respect to their DNA, it seems, that the mitochondrial genetic apparatus *in toto*, especially the translational part (ribosomes, ribosomal factors, tRNA and aminoacyl-tRNA synthetases) has been studied more extensively in microbial eucaryotes like *Neurospora* and yeast than in other cells.

The structure and function of nucleic acids and of the proteins engaged in replication, transcription and translation of mitochondrial DNA from *Neurospora* will be compared with other mitochondrial systems and with the nuclear-cytoplasmic system of the same cell.

## II. Structural and Functional Specificity of the Genetic Apparatus in Mitochondria

### 1. DNA

It appears to be a rule that all eucaryotic organisms contain mitochondrial or promitochondrial DNA, with the possible exception of certain "petite" mutants from yeast (GOLDRING et al., 1970; NAGLEY and LINNANE, 1970).

With respect to their mitochondrial DNA, all eucaryotic cells can be divided into two groups: Mitochondria from sea urchin and vertebrates including amphibians, birds and mammals contain predominantly circular DNA with an average contour length of 5 microns, corresponding to a molecular weight of $10 \times 10^6$ Daltons. Mitochondria from eucaryotic microorganisms like *Neurospora*, yeast or *Tetrahymena*, and from higher plants contain mostly linear DNA of a molecular weight larger than $20 \times 10^6$ Daltons. The average content of DNA per mitochondrion seems to be similar in both groups of organisms, namely ca. $7 \times 10^{-17}$ g or $40 \times 10^6$ Daltons (GRANICK and GIBOR, 1967; BORST et al., 1967; BORST and KROON, 1969; NASS, 1967; RABINOWITZ, 1968; NASS, 1969b).

Mitochondrial DNA has been isolated first by LUCK and REICH (1964) from *Neurospora*. These authors could identify mitochondrial DNA as a light satellite peak in a CsCl gradient, but their DNA preparation appeared heterogenous in the size distribution, with contour lengths ranging from 2 to 25 microns and an average length of 6.6 microns (corresponding to $12 \times 10^6$ Daltons).

We have isolated mitochondrial DNA from *Neurospora* using a method similar as described for the isolation of intact *mycoplasma* DNA (BODE and MOROWITZ, 1967). The resulting preparation consisted of linear molecules with a length of 25 microns (corresponding to $50 \times 10^6$ Daltons) and contained less than 10% smaller fragments; not a single circle was observed (SCHÄFER, GRANDI and KÜNTZEL, manuscript in preparation). A somewhat higher molecular weight ($60 \times 10^6$) for mitochondrial DNA from *Neurospora* has been calculated from renaturation data (WOOD and LUCK, 1970). The molecular weight of mitochondrial DNA from yeast is still controversial (NASS, 1969b; BILLHEIMER and AVERS, 1969). There is one recent report on the occurrence of 25 micron circles in yeast mitochondria (HOLLENBERG et al., 1969). These authors have also calculated the molecular weight from renaturation data to be $50 \times 10^6$. From these results one can tentatively conclude that mitochondrial DNA in *Neurospora* and yeast has a molecular weight of 50 to $60 \times 10^6$. Whether this DNA is linear or circular in its native form remains still open.

A general characteristic property of mitochondrial DNA is its high degree of homogeneity, which can be inferred from the sharpness of the DNA-band in a CsCl gradient and from the steepness of the melting curve, and its ability to renature much more rapidly than nuclear DNA (BORST, 1967).

In some organisms, expecially in microorganisms and plants, mitochondrial and nuclear DNA differ in their buoyant density and melting point, indicating a different G+C content. In the case of *Neurospora* the two DNA species can easily be separated in a CsCl gradient, the mitochondrial DNA banding at 1.701 g/cm³ and the nuclear DNA at 1.712 g/cm³ (WOOD and LUCK, 1970). Mitochondrial DNA from yeast has an exceptional low density of 1.685 g/cm³, corresponding to a G+C content of only 21% (TEWARI et al., 1966). This DNA seems to be drastically altered by cytoplasmic "petite" mutations, which lead in some cases to an almost total deletion of G+C containing sequences leaving behind "nonsense" material in form of alternating Poly d(AT) (BERNARDI et al., 1968; MEHROTRA and MAHLER, 1968). The relative high AT content of wild type mitochondrial DNA from yeast could possibly be attributed to a certain content of nonsense sequences, which would consequently reduce the information content of this DNA (BERNARDI and TIMASHEFF, 1970).

The structure and physical properties of supercoiled and nicked circular DNA and its circular and catenated oligomers in vertebrate cells have been studied in a large number of laboratories (NASS, 1969b). A discussion of these results would clearly exceed the scope of this article.

## 2. DNA- and RNA-Polymerase

The capability of intact mitochondria to incorporate *in vitro* deoxyribonucleoside triphosphates into DNA has been demonstrated by several authors (NEUBERT et al., 1968; PARSON and SIMPSON, 1968; WINTERSBERGER, 1968). Most authors agree that mitochondrial DNA synthesis is replicative rather than a repair process (SCHULTZ and NASS, 1967; MEYER and SIMPSON, 1968; KAROL and SIMPSON, 1968).

Mitochondrial DNA-Polymerase has been isolated and partially purified from rat liver (NEUBERT et al., 1967; MEYER and SIMPSON, 1968; KALF and CH'IH, 1968) and yeast (WINTERSBERGER and WINTERSBERGER, 1970a; IWASHIMA and RABINOWITZ, 1969). The results can be summarized as follows: Mitochondrial and nuclear DNA-Polymerase differ in molecular size, in their magnesium requirement and in their template dependence. Native mitochondrial DNA seems to be a better template for the mitochondrial than for the nuclear enzyme in yeast (WINTERSBERGER and WINTERSBERGER, 1970a). The rat liver mitochondrial DNA-Polymerase has been shown to prefer native mitochondrial DNA and is believed to produce replicates of mitochondrial DNA (KALF and CH'IH, 1968). However, it is clearly premature to decide whether mitochondrial DNA-Polymerase has repairing or replicating functions *in vivo*.

The presence of a DNA-dependent RNA-Polymerase in mitochondria can be inferred from the observation, that intact mitochondria are capable to incorporate ribonucleoside triphosphates into high molecular weight RNA (LUCK and REICH, 1964; WINTERSBERGER, 1964; WINTERSBERGER and TUPPY, 1965; WINTERSBERGER, 1966; NEUBERT et al., 1968; SACCONE et al., 1969). RNA synthesis in isolated mitochondria and mitochondrial membrane fragments from yeast is inhibited by actinomycin (WINTERSBERGER, 1964) but not by rifamycin (WINTERSBERGER and WINTERSBERGER, 1970b; HERZFELD, 1970), indicating that mitochondrial RNA-Polymerase may be similar to the nuclear enzymes, which are also rifamycin-insensitive (WEHRLI et al., 1968).

However, since all attempts to solubilize and to purify mitochondrial RNA-Polymerase have so far been unsuccessful, the interesting question remains still open whether the nuclear and mitochondrial enzymes are different or have different cofactors.

### 3. tRNA, Aminoacyl-tRNA-Synthetases and Transformylase

Mitochondria have been shown to contain transfer RNA and aminoacyl-tRNA-synthetases (WINTERSBERGER, 1966; FOURIER and SIMPSON, 1968; BARNETT and BROWN, 1967).

BARNETT et al. (1967) reported that purified mitochondria from *Neurospora* contained at least 15 different species of tRNA and their corresponding aminoacyl-tRNA-synthetases. They also found that mitochondrial aspartyl-, phenylalanyl- and leucyl-tRNA-synthetases could acylate only mitochondrial, but not cytoplasmic tRNA. Chromatographic differences were found between mitochondrial and cytoplasmic leucyl-, methionyl- and seryl-tRNA from *Neurospora* (BROWN and NOVELLI, 1968) and between mitochondrial and cytoplasmic leucyl-tRNA from *Tetrahymena* (SUYAMA and EYER, 1967). Mitochondrial and cytoplasmic leucyl-tRNA in *Neurospora* seem to recognize different codons (EPLER and BARNETT, 1967).

Similar results have been obtained in the case of rat liver mitochondria, which contain species of leucyl-, tyrosyl-, aspartyl-, valyl- and seryl-tRNAs exclusively located in mitochondria. It is interesting that these mitochondrial species cannot be acylated by cytoplasmic aminoacyl-tRNA-synthetases (BUCK and NASS, 1968, 1969).

One tRNA species should be discussed in more detail: It is now well established that not only procaryotic but also eucaryotic cells contain two methionyl-tRNA species, one of which can be formylated by bacterial transformylase. However, the presence of transformylase and hence of the polypeptide chain starter N-Formyl-methionyl-tRNA (fMet-tRNA) has longtime been believed to be restricted to bacteria (MARCKER and SMITH, 1969). More recently, fMet-tRNA has been detected also in mitochondria from yeast and rat liver (SMITH and MARCKER, 1968), in HeLa cells (GALPER and DARNELL, 1969) and in *Neurospora* (KÜNTZEL and SALA, 1969).

The specificity of aminoacylation and formylation of methionyl-tRNA from *E. coli* and *Neurospora* mitochondria and cytoplasm has been studied by

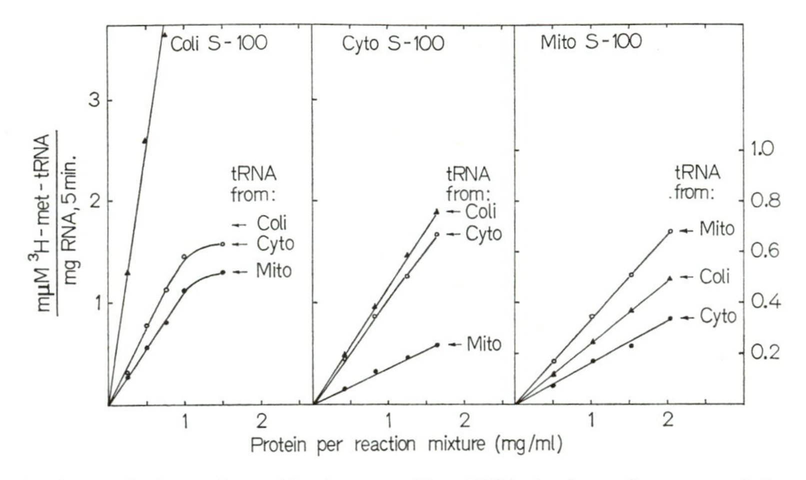


Fig. 1. Aminoacylation of methionine-specific tRNA in homologous and heterologous systems from *E. coli* and *Neurospora* mitochondria and cytoplasm. The reaction mixture (final volume 50 μl) contained (in μmoles per ml): Tris-HCl pH 7.5 (50), $MgCl_2$ (10), ATP (4), phosphoenolpyruvate (10), pyruvate kinase (50 μg), tRNA (3 mg), $^3$H-methionine (5 μC) und 100,000 × g supernatant protein as indicated. After incubation at 37° C for 5 minutes the cold TCA precipitable radioactivity was counted in a Tricarb

Table 1. *Synthesis of $^3$H-Methionyl-tRNA and N-Formyl-$^3$H-methionyl-tRNA in homologous and heterologous systems from E. coli, and Neurospora mitochondria and cytoplasm*

| tRNA | 100,000 × g supernatant | Radioactivity (%) | | |
|---|---|---|---|---|
| | | Met-A | fMet-A | Met |
| *E. coli* | *E. coli* | 33.3 | 65.4 | 1.3 |
| | *Neurospora*: mitochondria | 48.7 | 49.8 | 1.5 |
| | cytoplasm | 98.7 | 0.2 | 1.1 |
| *Neurospora*: | | | | |
| mitochondria | *E. coli* | 44.6 | 53.8 | 1.6 |
| | *Neurospora*: mitochondria | 41.5 | 56.6 | 1.9 |
| | cytoplasm | 98.2 | 0.1 | 1.7 |
| cytoplasm | *E. coli* | 61.0 | 36.8 | 2.2 |
| | *Neurospora*: mitochondria | 66.3 | 31.9 | 1.8 |
| | cytoplasm | 97.9 | 0.1 | 2.1 |

The incubation conditions were the same as described under Fig. 1, except that $N^{10}$-Formyl-tetrahydrofolic acid (10 mμM/ml) has been added. After 15 minutes incubation at 37° C, the tRNA was reisolated by phenol extraction at pH 5 and treated with RNAse. The adenosylesters were separated by electrophoresis at pH 3.5.

comparing the activity of homologous and heterologous systems (KÜNTZEL, manuscript in preparation). Fig. 1 shows the initial rate of aminoacylation of methionyl-tRNA as a function of the enzyme concentration. It is obvious that all three enzymes (from bacteria, mitochondria and cytoplasm) amino-

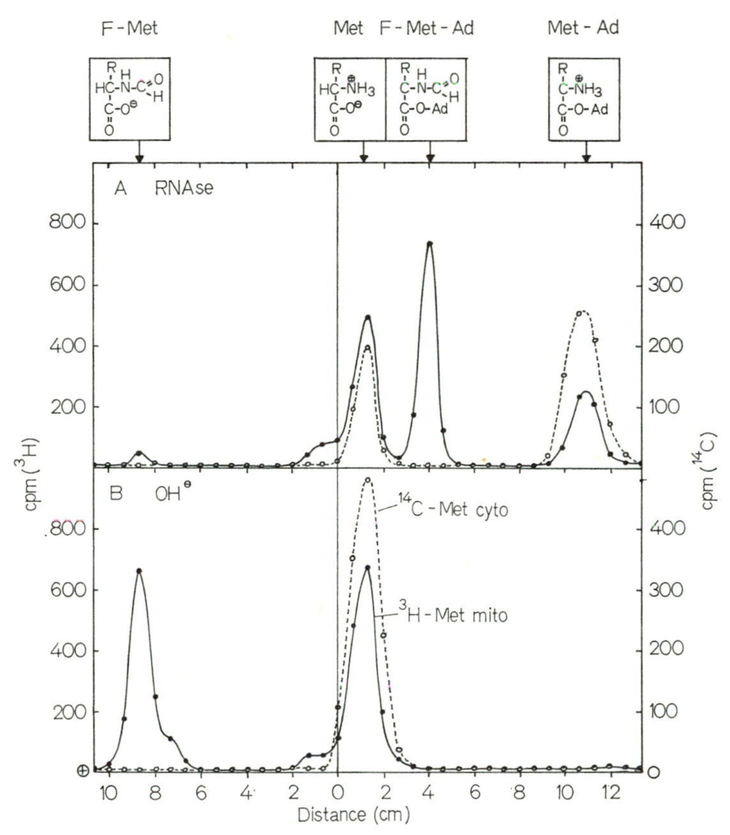


Fig. 2A and B. Electrophoresis of products obtained after RNAse (A) or alkali (B) treatment of mitochondrial N-Formyl-$^{3}$H-methionyl-tRNA and cytoplasmic $^{14}$C-methionyl-tRNA from *Neurospora*. Intact mitochondria were incubated with $^{3}$H-methionine, a 30,000 × g supernatant from cytoplasm with $^{14}$C-methionine, under conditions similar as described (KÜNTZEL, 1969a). The mitochondrial and cytoplasmic tRNA preparations obtained by phenol extraction were mixed, one aliquot was treated with RNAse (10 μg/ml, 5 minutes 37° C), another with diluted ammonia (pH 10, 2 hours at 37° C). The samples were subjected to electrophoresis at pH 3.5 (2 hours, 1,500 V), the dried pherograms were cut in stripes and counted in a Tricarb

acylate more rapidly the homologous than the heterologous tRNA, and, more surprisingly, that the *E. coli* system can be better combined with the cytoplasmic than with the mitochondrial system.

The transformylase activity of the same three enzyme preparations is compared in Table 1. The degree of formylation was measured by treating the tRNA with RNAse and separating the adenosyl-esters of methionine and N-formyl-methionine electrophoretically. Fig. 2 shows the electrophoretic pattern of the products obtained by treating mitochondrial fMet-tRNA with RNAse (A) or alkali (B). The data of Table 1 indicate that both *E. coli* and mitochondrial supernatants can formylate all three tRNA preparations, in contrast to the cytoplasmic supernatant, which does not contain transformylase activity to a measurable amount.

The implications of these findings for the peptide chain initiation mechanism in mitochondria and cytoplasm will be discussed later.

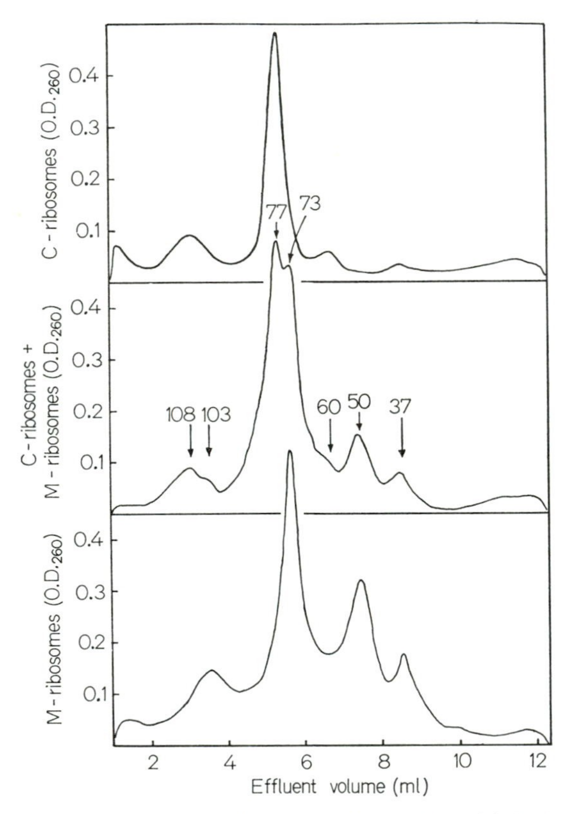


Fig. 3. Sedimentation pattern of cytoplasmic and mitochondrial ribosomes from *Neurospora*. Ribosomes were layered on exponential sucrose gradients containing 100 mM $NH_4Cl$, 10 mM $MgCl_2$ and 10 mM Tris pH 7.5 and centrifuged for 4 hours at 40,000 rpm in the Spinco SW 40 rotor at 2° C. The gradients were collected and recorded with a Gilford spectrophotometer equipped with a quartz flow cell. For further details see KÜNTZEL (1969b)

### 4. Ribosomes and Ribosomal Subunits

First evidence for the possible presence of ribosomes in mitochondria came from numerous morphological studies (for references see NASS, 1969). The ribosome-like particles shown in electromicrographs usually have a diameter of 120—150 Å and appear distinctly smaller than the cytoplasmic ribosomes measuring 180—200 Å. However, these observations were disputed by others, who could not find mitochondrial ribosomes differing in sedimentation constants or base composition from cytoplasmic ribosomes (TRUMAN, 1963; RABINOWITZ et al., 1966).

The existence of a mitochondrial class of ribosomes was proven by the isolation of mitochondrial ribosomes from *Neurospora* which were shown to be separable from cytoplasmic ribosomes in sucrose gradients (KÜNTZEL and NOLL, 1967; KÜNTZEL, 1969b). Fig. 3 shows the sedimentation pattern of mitochondrial 73*s* ribosomes, cytoplasmic 77*s* ribosomes and a mixture of

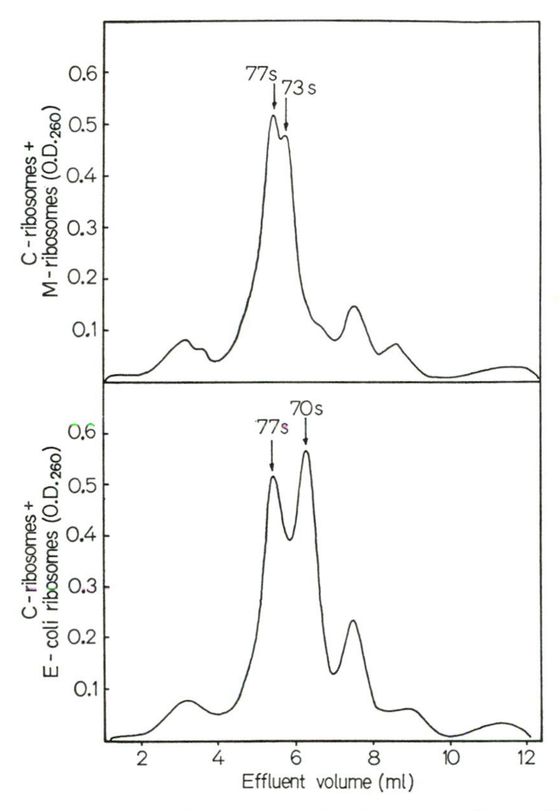


Fig. 4. Sedimentation pattern of cytoplasmic ribosomes from *Neurospora* and 70 *s* ribosomes from *E. coli*. Experimental details are described under Fig. 3

both particles. That mitochondrial ribosomes sediment faster than 70*s* ribosomes from *E. coli* is demonstrated in Fig. 4.

Mitochondrial ribosomes resemble bacterial ones in their dissociation properties: both particles are completely dissociated into subunits at 0.1 mM $MgCl_2$, whereas cytoplasmic ribosomes dissociate only after lowering the $MgCl_2$ concentration to 0.004 mM. The mitochondrial subunits sediment with 37*s* and 50*s* (compared with *E. coli* 30*s* and 50*s* particles as standards), the cytoplasmic subunits with 37*s* and 60*s* (KÜNTZEL, 1969b). Mitochondrial ribosomes have been isolated from *Neurospora* independently by RIFKIN et al. (1967). The fact that these authors find virtually the same base composition of mitochondrial rRNA, suggests that both groups are studying the same particle. However, RIFKIN et al. report higher *s*-values for both ribosomes from *Neurospora*; but this discrepancy should not be taken too serious because their reference ribosome, which is the familiar 70*s* ribosome from *E.coli*, sediments unusually fast with 81*s* under their conditions. Furthermore, a separation of the two *Neurospora* ribosomes is not possible under the gradient conditions used by RIFKIN et al.

A similar type of mitochondrial ribosomes sedimenting with 75 to 80*s* compared with 80*s* cytoplasmic ribosomes has been isolated from yeast (VIGNAIS et al., 1969; SCHMITT, 1969; MORIMOTO and HALVORSON, 1970; STEGEMAN et al., 1970).

As mentioned earlier, vertebrate mitochondria differ from microbial or plant mitochondria in their DNA, which is about five times smaller than yeast of *Neurospora* mitochondrial DNA. Another basic difference seems to be the smaller size of ribosomes and ribosomal RNA from vertebrate mitochondria. Traces of an unusually small particle sedimenting with 55*s* have been detected in rat liver by two groups (O'BRIEN and KALF, 1967; ASHWELL and WORK, 1970). ASHWELL's 55*s* particles contain nascent peptide chains which can be released in a chloramphenicol-sensitive step by puromycin, allowing the conclusion that the peptide might have been attached to a ribosome or a subunit.

However, one has to await evidence, such as for the presence of two subunits and two RNA molecules in this particle and its functional activity, to identify it as a ribosome. Convincing and conclusive evidence for the existence of a ribosome in vertebrate mitochondria comes from recent studies of SWANSON and DAVID (1970), who have isolated a 60*s* particle from *Xenopus laevis* mitochondria. This particle has been shown to contain two ribosomal RNA species (21*s* and 13*s*) which hybridize with mitochondrial DNA and which don't share any significant sequence homology with cytoplasmic rRNA. Furthermore, the 60*s* ribosome is active in a Poly U-dependent cell-free system. Two additional 43*s* and 32*s* particles found in *Xenopus* mitochondria are probably subunits of the 60*s* ribosome.

From these results it appears likely that rat liver 55*s* particles are mitochondrial ribosomes as well, and that all mitochondria containing 5 micron DNA circles have also "minimal" ribosomes sedimenting with 55*s* or 60*s*.

It will be of considerable interest to find out, whether this odd ribosome species exhibits the same specificity in chain initiation, elongation and antibiotic sensitivity as the bacterial ribosome.

### 5. Ribosomal RNA

Ribosomal RNA (rRNA) has been isolated from mitochondria or mitochondrial ribosomes of Yeast, *Neurospora* and *Aspergillus nidulans* (WINTERSBERGER, 1967; ROGERS et al., 1967; RIFKIN et al., 1967; KÜNTZEL and NOLL, 1967; DURE et al., 1967; FAUMAN and RABINOWITZ, 1969; EDELMAN et al., 1970).

Most workers find two main species sedimenting with 23*s* and 16*s* and in some cases an additional 13*s* peak. Thus, the mitochondrial rRNA differs clearly from cytoplasmic 25*s* and 17*s*, but resembles bacterial and chloroplast rRNA in the sedimentation constants (Table 2). On the other hand, mito-

Table 2. *Relative sedimentation coefficients of ribosomal RNA from three classes of ribosomes in the presence of* $Mg^{++}$ *or* $Na^{+}$

| Class of ribosomes | Source | rRNA components | | | | Ratio s (large)/ s (small) | |
|---|---|---|---|---|---|---|---|
| | | large | small | large | small | | |
| | | $Mg^{++}$ (1 mM) | | $Na^{+}$ (10 mM) | | $Mg^{++}$ | $Na^{+}$ |
| Animal cytoplasm | rat liver cytoplasm | 30.4 | 18.9 | 29.5 | 17.8 | 1.66 | 1.66 |
| Plant cytoplasm | bean cytoplasm | 26.5 | 16.0 | 24.7 | 17.0 | 1.65 | 1.45 |
| | *Neurospora* cytoplasm | 25.8 | 16.5 | 25.0 | 17.4 | 1.56 | 1.44 |
| Bacteria | *E. coli* | 22.6 | 16.0 | 21.0 | 16.0 | 1.41 | 1.31 |
| Organells | *Neurospora* mitochondria | 23.0 | 16.2 | 20.5 | 16.4 | 1.42 | 1.25 |
| | bean chloroplasts | 22.6 | 15.9 | 20.8 | 15.7 | 1.42 | 1.33 |

The data are from KÜNTZEL and NOLL (1967). The internal standard was 16.0*s* rRNA from *E. coli*.

Table 3. *Base composition of cytoplasmic and mitochondrial ribosomal RNA from Neurospora*

| | AMP | UMP | GMP | CMP | Uniden-tified |
|---|---|---|---|---|---|
| Cytoplasmic RNA: 25*s* + 17*s* | 24.1 ± 0.5 | 24.3 ± 0.8 | 28.1 ± 0.2 | 21.1 ± 0.3 | 2.3 ± 0.6 |
| Mitochondrial RNA: 21*s* + 16*s* | 27.2 ± 0.5 | 29.8 ± 0.7 | 22.9 ± 0.5 | 14.8 ± 0.8 | 5.7 ± 0.1 |
| Mitochondrial RNA: 14*s* + 10*s* | 29.7 ± 0.4 | 28.6 ± 0.5 | 20.4 ± 0.2 | 19.7 ± 0.4 | 1.6 ± 0.3 |

The data are from KÜNTZEL and NOLL (1967).

chondrial rRNA differs from both bacterial and cytoplasmic rRNA in its base composition, which is characterized by an unusually low G + C-content (Table 3).

EDELMAN et al. (1970) have determined the molecular weight of mitochondrial rRNA from *Aspergillus* by gel electrophoresis and reached the conclusion that mitochondrial rRNA is larger than bacterial rRNA in spite of the coincidence of *s*-values.

It seems, therefore, that mitochondrial rRNA is a distinct molecular species different from both bacterial and eucaryotic rRNA. The data concerning high molecular RNA in mammalian mitochondria are still confusing: KROON (1968) isolates 23*s* and 16*s* rRNA from rat liver mitochondria, DUBIN and BROWN (1967) find 27*s* and 18*s* RNA in Hamster cell mitochondria, VESCO and PENMAN (1969) and ATTARDI and ATTARDI (1969) describe a 21*s* and 12*s* RNA species from HeLa cell mitochondria but don't believe this material to be ribosomal RNA. A recent paper by SWANSON and DAVID (1970) who have isolated 21*s* and 13*s* RNA from functionally active 60*s* ribosomes of *Xenopus* mitochondria, has cleared the situation considerably: it seems now likely that the HeLa mitochondrial RNA species are indeed ribosomal, and that rat liver 55*s* particles contain similar rRNA species.

## 6. Ribosomal Proteins

The ribosomal proteins of mitochondrial ribosomes have been studied so far only in *Neurospora* (KÜNTZEL, 1969a). The proteins were labelled *in vivo* with

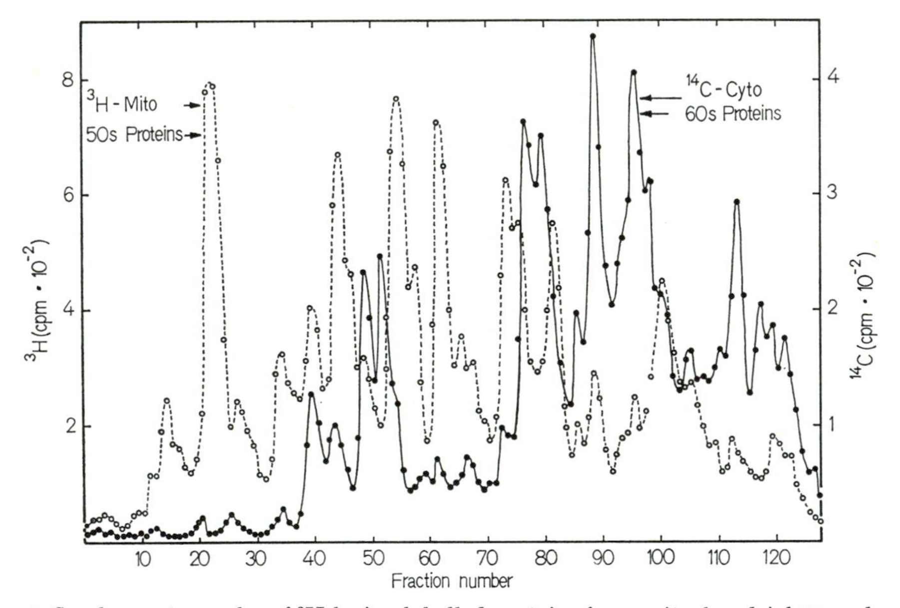


Fig. 5. Co-chromatography of $^{3}$H-lysine-labelled proteins from mitochondrial 50*s* subunits with $^{14}$C-lysine-labelled proteins from cytoplasmic 60 *s* subunits on a carboxymethyl cellulose column. For details see KÜNTZEL (1969a)

$^{3}$H- or $^{14}$C-lysine, the ribosomes were isolated from mitochondria and cytoplasm and dissociated into subunits. $^{14}$C-lysine labelled cytoplasmic 60*s* particles were mixed with $^{3}$H-lysine-labelled mitochondrial 50*s* subunits, treated with RNAse in presence of 6 M urea to digest the rRNA and chromatographed on a carboxymethyl cellulose column according to OTAKA et al. (1968).

Fig. 5 shows the elution profile of the proteins derived from the two big subunits. It is obvious that the cytoplasmic and mitochondrial subunits have almost no protein in common; the bulk of the cytoplasmic proteins elutes considerably later than most of the mitochondrial proteins. A similar result has been obtained with the two small subunits, which differ clearly in their protein pattern although they sediment at the same position in a sucrose gradient.

It seems therefore valid to conclude that, at least in *Neurospora*, mitochondrial ribosomes represent a unique class differing from both cytoplasmic 80*s* and bacterial 70*s* ribosomes. The structural similarities between bacterial and mitochondrial ribosomes (for example the *s*-values of their RNA-components and their large subunits) seem to be incidentical; however, the functional relatedness of the ribosome classes cannot be overlooked and will be discussed in the following sections.

## 7. Ribosomal Factors

The interaction of ribosomes with the peptide chain elongation factors G and T has been shown to be species-specific (for references see CIFERRI and PARISI, 1970).

It is, therefore, of considerable interest to know the specificity of the interaction of mitochondrial and cytoplasmic ribosomes with such factors. We have measured the Poly-U dependent polyphenylalanine synthesis in homologous and heterologous cell free system from *Neurospora* cytoplasm and mitochondria, from *E.coli* and rat liver (KÜNTZEL, 1969c). The results are shown in Table 4.

It is obvious that mitochondrial ribosomes can be combined with bacterial elongation factors and vice versa, whereas combinations of mitochondrial or bacterial ribosomes with supernatant enzymes from the cytoplasm of eucaryotic cells are not active. Even the two systems from the same cell exhibit a distinct incompatibility. The conclusion drawn from these data, that mitochondrial supernatant enzymes are specific for 70*s* ribosomes including mitochondrial 73*s* ribosomes, has recently been supported by the finding that the purified G-factor (translocase) from mitochondria can fully replace bacterial G-factor to complement with bacterial T-factor and 70*s* ribosomes (GRANDI and KÜNTZEL, 1970).

Another set of ribosomal factors specific for bacterial ribosomes are initiation factors recognizing fMet-tRNA; such factors have been found also in mitochondria from *Neurospora* (SALA and KÜNTZEL, 1970).

Mitochondrial ribosomes were separated into native subunits and 73*s* monosomes. These fractions were tested in their activity to bind fMet-tRNA and to synthesize fMet-puromycin. Table 5 shows that native subunits from mitochondria are active in both reactions; this activity can be removed by washing the ribosomes with 1 M $NH_4Cl$ and restored by adding initiation factors from *E.coli*. Mitochondrial 73*s* monosomes are less active but can be stimulated by *E.coli* factors. However, the cytoplasmic ribosomes from the same cell are not active, even in the presence of bacterial initiation factors.

Table 4. *Incorporation of ¹⁴C-phenylalanine from E. coli Phe-tRNA into polyphenylalanine in various cell free systems*

| Supernatant enzymes | Ribosomes | Incorporation of $^{14}$C-Phenylalanine | |
|---|---|---|---|
| | | μμmoles/mg RNA | % |
| *E. coli* | *E. coli* | 9.8 | 100 |
| | N.C. mitochondria | 9.2 | 94 |
| | N.C. cytoplasm | 0.2 | 2 |
| | rat liver cytoplasm | <0.1 | <1 |
| N.C. mitochondria | *E. coli* | 4.4 | 108 |
| | N.C. mitochondria | 4.1 | 100 |
| | N.C. cytoplasm | 0.9 | 23 |
| | rat liver cytoplasm | 0.8 | 19 |
| N.C. cytoplasm | *E. coli* | 0.1 | 1 |
| | N.C. mitochondria | 2.8 | 38 |
| | N.C. cytoplasm | 7.4 | 100 |
| | rat liver cytoplasm | 6.8 | 92 |
| Rat liver cytoplasm | *E. coli* | <0.1 | <1 |
| | N.C. mitochondria | 0.8 | 7 |
| | N.C. cytoplasm | 9.8 | 90 |
| | rat liver cytoplasm | 11.0 | 100 |

The data are from KÜNTZEL (1969c).

From these data we conclude that mitochondrial native subunits posess initiation factors which stimulate binding and translocation of fMet-tRNA. The mitochondrial factors resemble bacterial initiation factors, because they can replace each other. Cytoplasmic ribosomes from *Neurospora* may contain initiation factors as well, but they differ from bacterial or mitochondrial factors at least in their inability to recognize fMet-tRNA.

These results, together with the finding that mitochondria contain transformylase and incorporate N-formylmethionine into protein (KÜNTZEL and SALA, 1969), strongly support the view that mitochondria and bacteria share a common peptide chain initiation mechanism.

Finally, one other specific property common for mitochondrial and bacterial ribosomes should be mentioned; both particles are sensitive to chloramphenicol and resistent to cycloheximide, in contrast to cytoplasmic ribosomes from eucaryotes, which show the reverse response (KÜNTZEL, 1969a).

The similarity of bacteria and mitochondria has often been stressed (NASS, 1969a). There are, of course, structural analogies such as size and shape of bacteria and mitochondria, circularity of DNA, structure of membranes, sedimentation constants of ribosomes or ribosomal RNA etc., which cannot be overlooked. However, the "bacterial" character of mitochondria is probably

Table 5. *Binding of fMet-tRNA and synthesis of fMet-puromycin by different ribosomal preparations, in the presence and absence of initiation factors from E. coli*

| Ribosomes from | *E. coli* initiation factors | Binding of fMet-tRNA (μμmoles per mg ribosomes) | Synthesis of fMet-puromycin (μμmoles per mg ribosomes) |
|---|---|---|---|
| *Mitochondria of N. crassa* | | | |
| Native subunits: | | | |
| unwashed | — | 24.5 | 31.5 |
| washed | — | 1.7 | — |
| washed | + | 65.2 | — |
| 73 *s* ribosomes: | | | |
| unwashed | — | 1.5 | 0.9 |
| unwashed | + | 20.0 | 12.2 |
| *Cytoplasm of N. crassa* | | | |
| Unfractionated ribosomes: | | | |
| unwashed | — | 1.3 | 0.6 |
| unwashed | + | 2.7 | 1.0 |
| *E. coli* | | | |
| Unfractionated ribosomes: | | | |
| unwashed | — | 15.2 | 16.4 |
| unwashed | + | 43.8 | 64.8 |
| washed | — | 0.8 | 0.4 |
| washed | + | 49.7 | 45.9 |

The data are from SALA and KÜNTZEL (1970).

best documented by the presence of few key proteins like methionyl-tRNA transformylase, initiation factors recognizing fMet-tRNA, a translocase recognizing bacterial ribosomes and ribosomal proteins recognizing chloramphenicol.

Whether these proteins are coded by mitochondrial or nuclear genes is one of the most intriguing questions of mitochondrial biogenesis to be solved.

## III. Biosynthesis of the Genetic Apparatus

### 1. Replication and Proliferation of Mitochondrial DNA

Three hypothesis concerning the biogenesis and proliferation of mitochondria have been discussed for a long time (for references see SCHATZ, 1969):

1. Mitochondrial formation from other cell structures,
2. *De novo* formation of mitochondria, and
3. formation by growth and division of preexisting mitochondria and physical proliferation of complete mitochondria to the daughter cells.

Evidence favoring the third mechanism has since been accumulated, and one of the decisive experiments has been performed with exponentially growing

*Neurospora* cells by LUCK (1963). He and others could also show that mitochondrial DNA in *Neurospora* and yeast replicates in the classical semiconservative mechanism (REICH and LUCK, 1966; CORNEO et al., 1966; GROSS and RABINOWITZ, 1969). Mitochondrial division and replication of its DNA has been demonstrated to be periodic in synchronized *Neurospora* and yeast cells, the mitochondrial cycle being different from the mitotic cycle (HAWLEY and WAGNER, 1967; SMITH et al., 1968).

The presence of DNA and of a protein synthesizing apparatus in mitochondria does not in itself prove a genetic function of mitochondrial DNA. That such a genetic function not only exists but can also be transferred *via* mitochondria to daughter cells has been shown by DIACUMAKOS et al. (1965), who injected mitochondria from a cytoplasmic *Neurospora* "poky" mutant into wild type cells, thus transmitting the "poky" character, which results in a production of respiratory-deficient mitochondria.

While such results exclude a direct interaction between chromosomal and mitochondrial DNA, a nuclear control of the replication of mitochondrial DNA cannot be excluded. A regulation by a nuclear gene product would best explain why nuclear and mitochondrial DNA replication in *Neurospora* and yeast shows a constant shift of the periods (HAWLEY and WAGNER, 1967; SMITH et al., 1968). However, a direct control of mitochondrial replication by proteins of nuclear origin in yeast has been excluded by GROSSMAN et al. (1969), who observed a preferential synthesis of mitochondrial DNA in the absence of extramitochondrial protein synthesis.

While a nuclear control of mitochondrial replication remains likely, the possibility of a mitochondrial control of nuclear replication or cell division can be excluded, at least for yeast, by the finding that certain non-lethal neutral "petite" mutations lead to a complete loss of mitochondrial DNA (GOLDRING et al., 1970; NAGLEY and LINNANE, 1970).

## 2. Biosynthesis of Mitochondrial RNA

The study of genetic origin and biosynthesis of mitochondrial transfer-, ribosomal- and messenger-RNA *in vivo* would depend on the possibility to inhibit selectively mitochondrial or nuclear transcription. Such selective inhibitors have been found for mammalian cells (DUBIN, 1967; ZYBLER et al., 1969; VESCO and PENMAN, 1969) but not for *Neurospora* and yeast. All our knowledge about mitochondrial genes for stable RNA in the latter organisms depends, therefore, on hybridization data which are sometimes difficult to interprete.

WINTERSBERGER and VIEHHAUSER (1968) and FUKUHARA (1968) report that mitochondrial rRNA from yeast hybridizes both with mitochondrial and nuclear DNA, whereas cytoplasmic rRNA failes to hybridize with mitochondrial DNA. The hybridization between mitochondrial RNA and nuclear DNA is not due to sequence homologies between the two rRNA classes, because they do not compete with each other (WINTERSBERGER and VIEHHAUSER, 1968), but rather due to contamination of mitochondrial RNA by RNA of nuclear origin. Indeed, FUKUHARA (1970) has recently shown that RNA dehybridized from mitochondrial DNA does not hybridize with nuclear DNA. The number of genes for rRNA per mitochondrial genome in *Neurospora* has been estimated to be six for each 23 *s* and 16 *s* rRNA (WOOD and LUCK, 1969). A similar value based on a genome length of $60 \times 10^6$ Daltons can be calculated for yeast mitochondrial DNA from WINTERSBERGER's data (WINTERSBERGER, 1967).

The RNA gene products of vertebrate mitochondrial DNA seem to be 21 *s* and 12 *s* or 13 *s* RNA. This can be concluded from the finding that 21 *s* and 13 *s* RNA from HeLa cells and from *Xenopus* hybridizes exclusively with mitochondrial DNA (ZYLBER et al., 1969; SWANSON and DAVID, 1970), and that the *in vivo* biosynthesis of these species in HeLa cells is selectively inhibited by ethidium bromide, which is known to interact with circular mitochondrial DNA (VESCO and PENMAN, 1969).

Hybridization of aminoacyl-tRNA with mitochondrial DNA has been studied only in the rat liver system so far (NASS and BUCK, 1969). These authors found that various mitochondrial tRNA species hybridized much better with covalent circular mitochondrial DNA than their cytoplasmic counterparts.

That mitochondrial tRNA (4 *s* RNA) is indeed a transcription product of mitochondrial DNA can be followed from *in vivo* experiments with inhibitors, which block selectively mitochondrial or nuclear transcription (KNIGHT, 1969; VESCO and PENMAN, 1969; ZYLBER and PENMAN, 1969; DUBIN and MONTENECOURT, 1970).

AAIJ et al. (1970) have separated the complementary strands of rat liver mitochondrial DNA; they report that RNA synthesized in intact mitochondria *in vitro* hybridizes only with the heavy (G+C rich) strand, indicating that mitochondrial DNA might be transcribed asymmetrically not only *in vitro* but also *in vivo*.

### 3. Biosynthesis of Proteins Involved in the Expression of the Mitochondrial Genome

From the hybridization data reviewed above, it appears reasonable to assume that ribosomal RNA is synthesized within the mitochondrion. One would expect the protein components of the mitochondrial ribosome to be synthesized in the same compartment.

However, three laboratories have independently reported the surprising finding that the bulk of mitochondrial ribosomal proteins is synthesized on cytoplasmic ribosomes from *Neurospora* and yeast (KÜNTZEL, 1969a; NEUPERT et al., 1969a, b; DAVEY et al., 1969).

This can be followed from *in vivo* pulse labelling experiments with *Neurospora* in the presence of antibiotics. The data of Table 6 show that the incorporation of $^3$H-lysine into the proteins of both mitochondrial and cytoplasmic ribosomes is inhibited by cycloheximide and unaffected by chloramphenicol, indicating that most, if not all, mitochondrial ribosomal proteins

Table 6. *In vivo incorporation of $^{3}$H-lysine into mitochondrial and cytoplasmic ribosomal proteins in the presence and absence of antibiotics*

| Protein fraction | Cyclo-heximide | Chlor-amphenicol | cpm per mg protein | Inhibition (%) |
|---|---|---|---|---|
| Cytoplasmic ribosomes | — | — | 67,430 | — |
| | + | — | 2,090 | 96.9 |
| Mitochondrial ribosomes | — | — | 56,820 | — |
| | + | — | 1,940 | 97.2 |
| Cytoplasmic ribosomes | — | — | 91,500 | — |
| | — | + | 89,000 | 2.8 |
| Mitochondrial ribosomes | — | — | 85,820 | — |
| | — | + | 83,790 | 2.2 |

The data are from KÜNTZEL (1969a).

are synthesized on cytoplasmic ribosomes. Similar experiments with *Neurospora* cells have been reported by NEUPERT et al. (1969a, b) who could in addition show that nascent peptides labelled in isolated intact mitochondria could completely be released by puromycin, indicating that intact mitochondria cannot synthesize ribosomal proteins.

DAWEY et al. (1969) reached a similar conclusion from the observation, that yeast mitochondria isolated from cells which have been grown in the presence of chloramphenicol, have active ribosomes. However, this is not a rigid proof for the extramitochondrial biosynthesis of mitochondrial ribosomal proteins, because chloramphenicol does not suppress completely mitochondrial protein synthesis.

These inhibition experiments cannot give an answer to the question, whether the proteins of mitochondrial ribosomes are coded by nuclear or mitochondrial DNA. The second alternative has been suggested by LINNANE et al. from the finding that erythromycin resistance of mitochondrial protein synthesis is inherited extrachromosomally (LINNANE et al., 1968; THOMAS and WILKIE, 1968). However, it remains to be shown that this mutation has affected a ribosomal protein; a mutational alteration of ribosomal RNA could also lead to a resistance toward antibiotics. It will be, therefore, of considerable interest to identify a mitochondrial ribosomal protein altered by extra-chromosomal mutation, and to study its biosynthesis. An extramitochondrial synthesis of such a protein which should be expected from the incorporation studies mentioned above, would require a transport of mitochondrial messenger RNA into the cytoplasm, a translation of this messenger by cytoplasmic ribosomes and a transport of the product back into the mitochondrion. It is obvious that this mechanism would not be the most economical.

Little is known about the biosynthesis and genetic origin of other proteins involved in replication, transcription and translation of the mitochondrial genome.

The *in vivo* biosynthesis of DNA-polymerase from rat liver mitochondria has been studied by CH'IH and KALF (1969). The observation that cycloheximide inhibits the labelling of the enzyme, has been interpreted as an extramitochondrial synthesis of mitochondrial DNA-polymerase. However, since the enzyme was purified only partially, such data have to be met with caution. The same conclusion has been drawn by WINTERSBERGER and WINTERSBERGER (1970a) from the much more relevant observation that mitochondria from cytoplasmic "petite" mutants of yeast contain as much DNA-polymerase as wild type mitochondria, although the "petite" mutation leads to an almost complete loss of mitochondrial DNA and ribosomes (WINTERSBERGER, 1967a).

The problem of genetic origin and biosynthesis of enzymes involved in the gene expression could be solved by studying mutants with altered enzymes. Unfortunately, such mutants are not yet available, with the single exception of a *Neurospora* mutant described by GROSS et al. (1968). This mutant contains an altered cytoplasmic leucyl-tRNA-synthetase and is devoid of the mitochondria-specific leucyl-tRNA-synthetase. The interesting point is that obviously a single mutation of a nuclear gene has resulted in the alteration of the cytoplasmic and the abolition of the mitochondrial enzyme, indicating that both enzymes or at least a common subunit of both enzymes are coded by a nuclear gene.

The genetic origin of most proteins involved in mitochondrial protein biosynthesis, including the enzymes with bacterial specificity, is still obscure. An answer to this question will give a key to the understanding of mitochondrial autonomy and mitochondrial evolution.

## IV. The Products of the Genetic Apparatus

It is now well established that most of the mitochondrial proteins including the enzymes of the catabolic cycles, of the respiratory chain and oxydative phosphorylation, are synthesized in the cytoplasm under nuclear control (SCHATZ, 1969).

At the same time we know that mitochondria never arise *de novo*, because they need for their biogenesis the products of mitochondrial genes. The key role of the few mitochondrial gene products for the architecture of the mitochondrion, and hence for the function of the aerobic cell, is underlined by the fact that a complete second protein synthesizing machinery with some old-fashioned properties has been conserved through evolution; a machinery which is as complex as a bacterial genetic system, but which has to produce only some twenty proteins instead of several thousands.

We are in the puzzling situation to know more about this genetic apparatus than about its products. Again, most results concerning the identity of mitochondrial products have been obtained from *Neurospora* and yeast.

Intact mitochondria incorporate amino acids into insoluble proteins of the inner membrane (NEUPERT et al., 1967; SCHATZ, 1969). In the case of *Neuro-*

*spora* these proteins have been labelled either *in vivo* in the presence of cycloheximide or *in vitro* in isolated mitochondria, and separated by gel electrophoresis (SEBALD et al., 1968, 1969). Some of the labelled bands have been shown to be absent in cytoplasmic "poky" mutants, indicating that they are altered by a mitochondrial mutation (SEBALD et al., 1968).

The function of these membrane proteins is still unclear, but there is evidence for an association with cytochrome oxidase (BIRKMAYER et al., 1969). The same authors exclude the possibility that cytochrome oxidase itself might be a mitochondrial gene product, a view held by EDWARDS and WOODWARD (1969). Another possible functional role is the involvement in the protein complex which confers oligomycin sensitivity to ATPase. This can be followed from the observation that ATPase activity of mitochondria or promitochondria from yeast "petite" mutants is oligomycin-resistant, and that this resistance is inherited extrachromosomally (CRIDDLE and SCHATZ, 1969; SCHATZ and SALTZGRABER, 1969; WAKABAYASHI and GUNGE, 1970). However, the so called "oligomycin sensitivity conferring protein" (OSC-protein) of yeast mitochondria is not synthesized on mitochondrial ribosomes (TZAGOLOFF, 1970).

An interesting hypothesis that mitochondrial DNA might code for extramitochondrial membrane proteins has been forwarded by ATTARDI and ATTARDI (1967, 1968), who observed a preferential hybridization of messenger RNA derived from membrane-bound cytoplasmic polysomes with mitochondrial DNA from HeLa cells. Unfortunately, they could not exclude a contamination of their endoplasmic reticulum preparation with mitochondrial fragments containing mitochondrial RNA. However, a general role of mitochondrial DNA for the biosynthesis of membrane proteins is a tempting idea which should be tested more rigorously.

As a conclusion we have to state that our knowledge of number, molecular weight, and function of mitochondrial gene products is poor.

The possible number of mitochondrial genes can be roughly estimated from the molecular weight of mitochondrial DNA.

The genome length of vertebrate mitochondrial DNA has been estimated from renaturation data to be equivalent to its molecular weight (HOLLENBERG et al., 1969); this means that all 5 micron circles within one mitochondrion carry the same information equivalent to $10 \times 10^6$ Daltons or 15,000 base pairs. If we assume one gene per each of 20 tRNA's, we have to substract 1,500 base pairs. One copy of each 21 *s* and 13 *s* ribosomal RNA would require 3,000 base pairs as a minimum number (the molecular weight of 21 + 13 *s* RNA assumed to be ca. $1 \times 10^6$). The remaining genome would contain 10,500 base pairs which can code for 3,500 amino acids or 21 proteins of molecular weight 20,000. This is an upper number, because a possible redundancy of genes for tRNA, rRNA or protein would reduce the gene number.

To estimate the gene number per *Neurospora* or yeast mitochondrial genome we substract from total 75,000 base pairs (corresponding to $50 \times 10^6$ Daltons) the 1,500 base pairs coding for 20 tRNA's, and 22,500 base pairs coding for 6 copies of 23*s* and 16*s* rRNA. The remaining 51,000 base pairs or 17,000 triplets would code for 102 proteins having 20,000 Daltons. Again this number is probably smaller because of a certain degree of redundancy observed with mitochondrial DNA from *Neurospora* (Wood and Luck, 1969), and of a possible content of nonsense sequences (Mehrotra and Mahler, 1968).

The number of 102 protein genes calculated for *Neurospora* or yeast mitochondrial DNA would be sufficient to code for all ribosomal proteins (ca. 60 species) of the mitochondrial ribosome and for most of the other proteins involved in the genetic expression.

However, the number of 21 protein genes for vertebrate mitochondrial DNA is clearly too small to code for all ribosomal proteins of the 60*s* ribosome (the number can be estimated with 40 copies) and for most of the other proteins involved in mitochondrial protein synthesis.

Whatever the products of mitochondrial DNA will turn out to be, the genetic apparatus of mitochondria is a fascinating example of a "minimalized" and highly specialized system. The small size of mitochondrial DNA, but also the mystery of its informational role, makes this DNA an ideal subject for a complete *in vitro* transcription and translation of a cellular DNA.

## References

Aaij, C., Saccone, C., Borst, P., Gadaleta, M. N.: Hybridization studies with RNA synthesized by isolated rat liver mitochondria. Biochim. biophys. Acta (Amst.) **199**, 373—381 (1970).

Altmann, R.: Die Elementarorganismen und ihre Beziehungen zu den Zellen. Leipzig: Veit und Co. 1890.

Ashwell, M. A., Work, T. S.: The functional characterization of ribosomes from rat liver mitochondria. Biochem. biophys. Res. Commun. **39**, 204—211 (1970).

Attardi, B., Attardi, G.: A membrane-associated RNA of cytoplasmic origin in HeLa cells. Proc. nat. Acad. Sci. (Wash.) **58**, 1051—1058 (1967).

— — Sedimentation properties of RNA species homologous to mitochondrial DNA in HeLa cells. Nature (Lond.) **224**, 1079—1083 (1969).

Barnett, W. E., Brown, D. H.: Mitochondrial transfer ribonucleic acids. Proc. nat. Acad. Sci. (Wash.) **57**, 452—458 (1967).

— — Epler, J. L.: Mitochondrial-specific aminoacyl-RNA synthetases. Proc. nat. Acad. Sci. (Wash.) **57**, 1775—1781 (1967).

Bernardi, G., Carnevali, I., Nicolaieff, A., Piperno, G., Tecce, G.: Separation and characterization of a satellite DNA from a yeast cytoplasmic "petite" mutant. J. molec. Biol. **37**, 493—505 (1968).

— Timasheff, S. N.: Optical rotatory dispersion (OR) and circular dichroism (CD) properties of yeast mitochondrial DNA. J. molec. Biol. **48**, 43—52 (1970).

Billheimer, F. E., Avers, C. J.: Nuclear and mitochondrial DNA from wild type and petite yeast: circularity, length, and buoyant density. Proc. nat. Acad. Sci. (Wash.) **64**, 739—746 (1969).

Birkmayer, G. D., Sebald, W., Bücher, T.: Cytochrom-Oxidase und ein an diese assoziiertes, markiertes Protein aus $^{14}C$-markierten Mitochondrien von *Neurospora crassa*. Hoppe-Seylers Z. physiol. Chem. **350**, 1159 (1969).

Bode, H. R., Morowitz, H. J.: Size and structure of the *Mycoplasma hominis* H39 chromosome. J. molec. Biol. **23**, 191—199 (1967).

BORST, P., KROON, A. M.: Mitochondrial DNA: physicochemical properties, replication, and genetic function. Int. Rev. Cytol. **26**, 108—190 (1969).
— — RUTTENBERG, G. J. C. M.: Mitochondrial DNA and other forms of cytoplasmic DNA. In: Genetic elements, properties and function. New York: Academic Press 1967.
BROWN, D. H., NOVELLI, G. D.: Chromatographic differences between the cytoplasmic and mitochondrial tRNAs of *Neurospora crassa*. Biochem. biophys. Res. Commun. **31**, 262—266 (1968).
BUCK, C. A., NASS, M. M. K.: Differences between mitochondrial and cytoplasmic tRNA and aminoacyl-tRNA synthetases from rat liver. Proc. nat. Acad. Sci. (Wash.) **60**, 1045—1052 (1968).
— — Studies on mitochondrial tRNA from animal cells. I. A comparison of mitochondrial and cytoplasmic tRNA and aminoacyl-tRNA synthetases. J. molec. Biol. **41**, 67—82 (1969).
CHÈVREMONT, M., CHÈVREMONT-COMHAIRE, S., BAECKELAND, E.: Action de désoxyribonucléases neutre et acide sur des cellules somatiques vivantes cultivées *in vitro*. Arch. Biol. (Liège) **70**, 811—831 (1959).
CH'IH, J. J., KALF, G. F.: Studies on the biosynthesis of the DNA-polymerase of rat liver mitochondria. Arch. Biochem. **133**, 38—45 (1969).
CIFERRI, O., PARISI, B.: Ribosome specificity of protein synthesis *in vitro*. Progr. Nucl. Acid. Res. Mol. Biol. (in press).
CORNEO, G., MOORE, C., SANADI, D. R., GROSSMAN, L. I., MARMUR, J.: Mitochondrial DNA in yeast and in some mammalian species. Science **151**, 687—689 (1966).
CRIDDLE, R. S., SCHATZ, G.: Promitochondria of anaerobically grown yeast. I. Isolation and biochemical properties. Biochemistry **8**, 323—334 (1969).
DAVEY, P. J., YU, R., LINNANE, A. W.: The intracellular site of formation of the mitochondrial protein synthetic system. Biochem. biophys. Res. Commun. **36**, 30—34 (1969).
DIACUMAKOS, E. G., GARNJOBST, L., TATUM, E. L.: A cytoplasmic character in *Neurospora crassa*. The role of nuclei and mitochondria. J. Cell Biol. **26**, 427—443 (1965).
DUBIN, D. T.: The effect of actinomycin on the synthesis of mitochondrial RNA in hamster cells. Biochem. biophys. Res. Commun. **29**, 655—660 (1967).
— BROWN, R. E.: A novel ribosomal RNA in hamster cell mitochondria. Biochem. biophys. Acta (Amst.) **145**, 538—540 (1967).
— MONTENECOURT, B. S.: Mitochondrial RNA from cultured animal cells. Destinctive high molecular weight and 4 *s* species. J. molec. Biol. **48**, 279—295 (1970).
DURE, L. S., EPLER, J. L., BARNETT, E.: Sedimentation properties of mitochondrial and cytoplasmic ribosomal RNA's from *Neurospora crassa*. Proc. nat. Acad. Sci. (Wash.) **58**, 1883—1887 (1967).
EDELMAN, M., VERMA, J. M., LITTAUER, U. Z.: Mitochondrial rRNA from *Aspergillus nidulans:* characterization of a novel molecular species. J. molec. Biol. **49**, 67—83 (1970).
EDWARDS, D. L., WOODWARD, D. O.: An altered cytochrome oxidase in a cytoplasmic mutant of *Neurospora*. FEBS-letters **4**, 193—196 (1969).
EPLER, J. L., BARNETT, W. E.: Coding properties of *Neurospora* mitochondrial and cytoplasmic leucine tRNAs. Biochem. biophys. Res. Commun. **28**, 328—333 (1967).
FAUMAN, M., RABINOWITZ, M., GETZ, G. S.: Base composition and sedimentation properties of mitochondrial RNA of *Saccharomyces cerevisiae*. Biochim. biophys. Acta (Amst.) **182**, 355—360 (1969).
FOURNIER, M. J., SIMPSON, M. V.: The occurrence of amino acid activating enzymes and sRNA in mitochondria. In: Biochemical aspects of the biogenesis of mitochondria. Bari: Adriatica Editrice 1968.

FUKUHARA, H.: Informational role of yeast mitochondrial DNA studied by hybridization with different classes of RNA. In: Biochemical aspects of the biogenesis of mitochondria. Bari: Adriatica Editrice 1968.

— Transcriptional origin of RNA in a mitochondrial fraction of yeast and its bearing on the problem of sequence homology between mitochondrial and nuclear DNA. Molec. Gen. Genetics **107**, 58—70 (1970).

GALPER, J. B., DARNELL, J. E.: The presence of N-formyl-methionyl-tRNA in HeLa cell mitochondria. Biochem. biophys. Res. Commun. **34**, 205—214 (1969).

GIBOR, A., GRANICK, S.: Plastids and mitochondria: Inheritable systems. Science **145**, 890—897 (1964).

GOLDRING, E. S., GROSSMAN, L. I., KRUPNICK, D., CRYER, D. R., MARMUR, J.: The ethidium bromide (EB) induced breakdown of yeast mitochondrial DNA (mDNA) during induction of petites. Fed. Proc. **29**, 2710 (1970).

GRANDI, M., KÜNTZEL, H.: Mitochondrial peptide chain elongation factors from *Neurospora crassa*. FEBS-letters **10**, 25 (1970).

GRANICK, S., GIBOR, A.: The DNA of chloroplasts, mitochondria and centrioles. Progr. Nucl. Acid Res. Mol. Biol. **6**, 143—187 (1967).

GROSS, N. J., RABINOWITZ, J. B. C.: Synthesis of new strands of mitochondrial and nuclear DNA by semiconservative replication. J. biol. Chem. **244**, 1563—1566 (1969).

GROSS, S. R., MCCOY, M. T., GILMORE, E. B.: Evidence for the involvement of a nuclear gene in the production of the mitochondrial leucyl-tRNA synthetase of *Neurospora*. Proc. nat. Acad. Sci. (Wash.) **61**, 253—260 (1968).

GROSSMAN, L. I., GOLDRING, E. S., MARMUR, J.: Preferential synthesis of yeast mitochondrial DNA in the absence of protein synthesis. J. molec. Biol. **46**, 367—376 (1969).

HAWLEY, E. S., WAGNER, R. P.: Synchronous mitochondrial division in *Neurospora crassa*. J. Cell Biol. **35**, 489—499 (1967).

HERZFELD, F.: Rifampicin insensitivity of RNA synthesis in *Neurospora* mitochondria Hoppe-Seylers Z. physiol. Chem. **351**, 658—660 (1970).

HOLLENBERG, C. P., BORST, P., THURING, R. W. J., BRUGGEN, E. F. J., VAN: Size, structure and genetic complexity of yeast mitochondrial DNA. Biochim. biophys. Acta (Amst.) **186**, 417—419 (1969).

IWASHIMA, A., RABINOWITZ, M.: Partial purification of mitochondrial and supernatant DNA polymerase from *Saccharomyces cerevisiae*. Biochim. biophys. Acta (Amst.) **178**, 283—293 (1969).

KALF, G. F., CH'IH, J. J.: Purification and properties of DNA polymerase from rat liver mitochondria. J. biol. Chem. **243**, 4904—4921 (1968).

KAROL, M. H., SIMPSON, M. V.: DNA synthesis by isolated mitochondria: A replicative rather than a repair process. Science **162**, 470—473 (1968).

KNIGHT, E.: Mitochondrial RNA of HeLa cells. Effect of ethidium bromide on the synthesis of ribosomal and 4 *s* RNA. Biochemistry **8**, 5089—5093 (1969).

KROON, A. M.: RNA of rat liver mitochondrial ribosomes. 5th meeting Fed. Europ. Biochem. Socs., Prague 1968, Abstr. Nr 204.

KÜNTZEL, H.: The proteins of mitochondrial and cytoplasmic ribosomes from *Neurospora crassa*. Nature (Lond.) **222**, 142—146 (1969a).

— Mitochondrial and cytoplasmic ribosomes from *Neurospora crassa*: characterization of their subunits. J. molec. Biol. **40**, 315—320 (1969b).

— Specificity of mitochondrial and cytoplasmic ribosomes from *Neurospora crassa* in poly-U dependent cell free systems. FEBS-letters **4**, 140—142 (1969c).

— NOLL, H.: Mitochondrial and cytoplasmic polysomes from *Neurospora crassa*. Nature (Lond.) **215**, 1340—1345 (1967).

— SALA, F.: Kettenanfangsmechanismus der mitochondrialen Proteinbiosynthese. Hoppe-Seylers Z. physiol. Chem. **350**, 1158 (1969).

LINNANE, A. W., SAUNDERS, G. W., GINGOLD, E. B., LUKINS, H. B.: The biogenesis of mitochondria, V. Cytoplasmic inheritance of erythromycin resistance in *Saccharomyces cerevisiae*. Proc. nat. Acad. Sci. (Wash.) **59**, 903—910 (1968).

LUCK, D. J. L.: Formation of mitochondria in *Neurospora crassa*. A quantitative radioautographic study. J. Cell Biol. **16**, 483—499 (1963).

— REICH, E.: DNA in mitochondria of *Neurospora crassa*. Proc. nat. Acad. Sci. (Wash.) **52**, 931—938 (1964).

MARCKER, K. A., SMITH, A. E.: On the universality of the mechanism of polypeptide chain initiation. Bull. Soc. Chim. biol. (Paris) **51**, 1453—1458 (1969).

MEHROTRA, B. D., MAHLER, H. R.: Characterization of some unusual DNA's from the mitochondria from certain "peptide" strains of *Saccharomyces cerevisiae*. Arch. Biochem. **128**, 685—703 (1968).

MEYER, R. R., SIMPSON, M. V.: DNA biosynthesis in mitochondria: Partial purification of a distinct DNA polymerase from isolated rat liver mitochondria. Proc. nat. Acad. Sci. (Wash.) **61**, 130—137 (1968).

MORIMOTO, H., HALVORSON, H.: Characterization of yeast mitochondrial ribosomes. Fed. Proc. **29**, 2003 (1970).

NAGLEY, P., LINNANE, A. W.: Mitochondrial DNA deficient petite mutants of yeast. Biochem. biophys. Res. Commun. **39**, 989—996 (1970).

NASS, M. M. K.: Circularity and other properties of mitochondrial DNA of animal cells. In: Organizational biosynthesis. New York: Academic Press 1967.

— Mitochondrial DNA: advances, problems, and goals. Science **165**, 25—35 (1969a).

— BUCK, C. A.: Comparative hybridization of mitochondrial and cytoplasmic aminoacyl transfer RNA with mitochondrial DNA from rat liver. Proc. nat. Acad. Sci. (Wash.) **62**, 506—513 (1969).

— NASS, S.: Fibrous structures within the matrix of developing chick embryo mitochondria. Exp. Cell Res. **26**, 424—437 (1962).

NASS, S.: The significance of the structural and functional similarities of bacteria and mitochondria. Int. Rev. Cytol. **25**, 55—129 (1969b).

NEUBERT, D., HELGE, H., MERCKER, H.-J.: Biosynthesis of mammalian mitochondrial RNA. In: Biochemical aspects of the biogenesis of mitochondria. Bari: Adriatica Editrice 1968.

— OBERDISSE, E., SCHMIEDER, M., REINSCH, J.: Solubilization and some properties of vertebrate mitochondrial DNA polymerase. Hoppe-Seylers Z. physiol. Chem. **348**, 1709—1711 (1967).

NEUPERT, W., BRDICZKA, D., BÜCHER, T.: Incorporation of amino acids into the outer and inner membrane of isolated rat liver mitochondria. Biochem. biophys. Res. Commun. **27**, 488—493 (1967).

— SEBALD, W., SCHWAB, A. J., MASSINGER, P., BÜCHER, T.: Incorporation *in vivo* of $^{14}C$-labelled amino acids into the proteins of mitochondrial ribosomes from *Neurospora crassa* sensitive to cycloheximide and insensitive to chloramphenicol. Europ. J. Biochem. **10**, 589—591 (1969).

— — — PFALLER, A., BÜCHER, T.: Puromycin sensitivity of ribosomal label after incorporation of $^{14}C$-labelled amino acids into isolated mitochondria from *Neurospora crassa*. Europ. J. Biochem. **10**, 585—588 (1969a).

O'BRIEN, T. W., KALF, G. F.: Ribosomes from rat liver mitochondria. I. Isolation procedure and contamination studies. II. Partial characterization. J. biol. Chem. **242**, 2172—2179, 2180—2185 (1967).

OTAKA, E., ITOH, T., OSAWA, S.: Ribosomal proteins of bacterial cells: strain and species specificity. J. molec. Biol. **33**, 93—107 (1968).

PARSONS, P., SIMPSON, M. V.: Studies on DNA biosynthesis in isolated rat liver mitochondria. In: Biochemical aspects of the biogenesis of mitochondria. Bari: Adriatica Editrice 1968.

RABINOWITZ, M.: Extranuclear DNA. Bull. Soc. Chim. biol. (Paris) **50**, 311—349 (1968).
— DE SALLE, L., SINCLAIR, J., STIREWALT, R., SWIFT, H.: Ribosomes isolated from rat liver mitochondria preparation. Fed. Proc. **25**, 581 (1966).
REICH, E., LUCK, D. J. L.: Replication and inheritance of mitochondrial DNA. Proc. nat. Acad. Sci. (Wash.) **55**, 1600—1608 (1966).
RIFKIN, M. R., WOOD, D. D., LUCK, D. J. L.: Ribosomal RNA and ribosomes from mitochondria of *Neurospora crassa*. Proc. nat. Acad. Sci. (Wash.) **58**, 1025—1032 (1967).
ROGERS, P. J., PRESTON, B. N., TITCHENER, E. B., LINNANE, A. W.: Differences between the sedimentation characteristics of the RNA's prepared from yeast cytoplasmic ribosomes and mitochondria. Biochem. biophys. Res. Commun. **27**, 405—411 (1967).
ROODYN, D. B., WILKIE, D.: The biogenesis of mitochondria. London: Methuen 1968.
SACCONE, C., GADALETA, M. N., GALLERANI, R.: RNA synthesis in isolated rat liver mitochondria. Europ. J. Biochem. **10**, 61—65 (1969).
SALA, F., KÜNTZEL, H.: Peptide chain initiation in homologous and heterologous systems from mitochondria and bacteria. Europ. J. Biochem. **15**, 280—286 (1970).
SCHATZ, G.: Biogenesis of mitochondria. In: Membranes of mitochondria and chloroplasts. New York: Nostrand Reinhold Comp. 1970.
— SALTZGRABER: Protein biosynthesis by yeast promitochondria *in vivo*. Biochem. biophys. Res. Commun. **37**, 996—1001 (1969).
SCHMITT, H.: Characterization of mitochondrial ribosomes from *Saccharomyces cerevisiae*. FEBS-letters **4**, 234—238 (1969).
SCHULTZ, S. R., NASS, S.: DNA-dependent incorporation of tritiated thymidine triphosphate by a rat liver mitochondrial extract. J. Cell Biol. **35**, 123 A (1967).
SEBALD, W., BÜCHER, T., OLBRICH, B., KAUDEWITZ, F.: Electrophoretic pattern of and amino acid incorporation *in vitro* into the insoluble mitochondrial protein of *Neurospora crassa* wild type and mi-1 mutant. FEBS-letters **1**, 235—240 (1968).
— SCHWAB, A. J., BÜCHER, T.: Cycloheximide resistant amino acid incorporation into mitochondrial protein from *Neurospora crassa in vivo*. FEBS-letters **4**, 243—246 (1969).
SMITH, A. E., MARCKER, K. A.: N-formylmethionyl transfer RNA in mitochondria from yeast and rat liver. J. molec. Biol. **38**, 241—243 (1968).
SMITH, D., TAURO, P., SCHWEIZER, E., HALVORSON, H. O.: The replication of mitochondrial DNA during the cell cycle in *Saccharomyces lactis*. Proc. nat. Acad. Sci. (Wash.) **60**, 936—942 (1968).
STEGEMAN, W. J., COOPER, C. S., AVERS, C. J.: Physical characterization of ribosomes from purified mitochondria of yeast. Biochem. biophys. Res. Commun. **39**, 69—76 (1970).
SUYAMA, Y., EYER, J.: Leucyl tRNA and leucyl tRNA synthetase in mitochondria of *Tetrahymena pyriformis*. Biochem. biophys. Res. Commun. **28**, 746—751 (1967).
SWANSON, R. F., DAWID, I. B.: The mitochondrial ribosome of *Xenopus laevis*. Proc. nat. Acad. Sci. (Wash.) **66**, 117—124 (1970).
TEWARI, K. K., VÖTSCH, W., MAHLER, H. R.: Biochemical correlates of respiratory deficiency. V. Mitochondrial DNA. J. molec. Biol. **20**, 453—481 (1966).
THOMAS, D. Y., WILKIE, D.: Recombination of mitochondrial drug resistance factors in *Saccharomyces cerevisiae*. Biochem. biophys. Res. Commun. **30**, 368—372 (1968).
TRUMAN, D. E. S.: Incorporation of amino acids into the proteins of sub-mitochondrial particles. Exp. Cell Res. **31**, 313—320 (1963).
TZAGOLOFF, A.: Assembly of the mitochondrial membrane system. III. Function and synthesis of the oligomycin sensitivity conferring protein of yeast mitochondria. J. biol. Chem. **245**, 1545—1551 (1970).

VESCO, C., PENMAN, S.: The cytoplasmic RNA of HeLa cells: New discrete species associated with mitochondria. Proc. nat. Acad. Sci. (Wash.) **62**, 218—225 (1969).

— — Insensitivity of mitochondrial RNA synthesis to mengovirus infection in CHO cells. Nature (Lond.) **224**, 1021—1023 (1969).

VIGNAIS, P. V., HUET, J., ANDRE, J.: Isolation and characterization of ribosomes from yeast mitochondria. FEBS-letters **3**, 177—181 (1969).

WAGNER, R. P.: Genetics and phenogenetics of mitochondria. Science **163**, 1026—1031 (1969).

WAKABAYASHI, K., GUNGE, N.: Extrachromosomal inheritance of oligomycin resistance in yeast. FEBS-letters **6**, 302—304 (1970).

WEHRLI, W., NUESCH, J., KNÜSEL, F., STAEHELIN, M.: Action of rifamycine on RNA polymerase. Biochim. biophys. Acta (Amst.) **157**, 215—217 (1968).

WILKIE, D.: The cytoplasm in heredity. London: Methuen 1964.

WINTERSBERGER, E.: DNA-abhängige RNA-Synthese in Rattenleber-Mitochondrien. Hoppe-Seylers Z. physiol. Chem. **336**, 285—288 (1964).

— Synthesis and function of mitochondrial RNA. In: Regulation of metabolic processes in mitochondria. New York: Elsevier 1966.

— A distinct class of rRNA components in yeast mitochondria as revealed by gradient centrifugation and by DNA-RNA-hybridization. Hoppe-Seylers Z. physiol. Chem. **348**, 1701—1704 (1967).

— Synthesis of DNA in isolated yeast mitochondria. In: Biochemical aspects of the biogenesis of mitochondria. Bari: Adriatica Editrice 1968.

— TUPPY, H.: DNA-abhängige RNA-Synthese in isolierten Hefe-Mitochondrien. Biochem. Z. **341**, 399—408 (1965).

— VEHHAUSER, G.: Function of mitochondrial DNA in yeast. Nature (Lond.) **220**, 699—702 (1969).

— WINTERSBERGER, U.: Rifamycin insensitivity of RNA synthesis in yeast. FEBS-letters **6**, 58—60 (1970a).

WINTERSBERGER, U., WINTERSBERGER, E.: Studies on DNA-polymerases from yeast. 2. Partial purification and characterization of mitochondrial DNA-polymerase from wild type and respiratory-deficient yeast cells. Europ. J. Biochem. **13**, 20—27 (1970b).

WOOD, D. D., LUCK, D. J. C.: Hybridization of mitochondrial ribosomal RNA. J. molec. Biol. **41**, 211—224 (1969).

ZYLBER, E., PENMAN, S.: Mitochondrial-associated 4 *s* RNA synthesis inhibition by ethidium bromide. J. molec. Biol. **46**, 201—204 (1969).

— VESCO, C., PENMAN, S.: Selective inhibition of the synthesis of mitochondria-associated RNA by ethidium bromide. J. molec. Biol. **44**, 195—204 (1969).

Escuela Nacional de Ciencias Biológical del I. P. N.
México 17, D. F. México

# Effects of Freeze-Drying and Sporulation on Microbial Variation

MANUEL SERVIN-MASSIEU

With 6 Figures

Contents



They, too, swerved from their course; and, entering the Bayou of Plaquemine
Soon were lost in a maze of sluggish and devious waters
Which, like a network of steel, extended in every direction

Longfellow
from Evangeline

## I. Introduction

In the last few years we have seen the birth of several hybrid scientific disciplines like biophysics, molecular genetics, biocybernetics, bionics, etc., as a consequence of the achievements of geneticists, biochemists, virologists, classical biologists, physicists and even engineers (POLLARD, 1965) interested in the study of biological problems from different perspectives. This situation, in itself, establishes a paradox because the continuous growth of science introduces an ever increasing number of questions that again renew the mixing of scientific disciplines (HOLTON, 1962). Since a great deal of effort has been put for a long time into the preservation of microbial cultures and vaccines by cryodesiccation, we now find that one of the new sciences, cryobiology or the study of the activities of living organisms as influenced by low temperatures and freezing, has started to share spheres of influence with microbiology. Both fields then, share problems of mutual interest one of which is covered by the present review; knowledge of the various aspects of this problem is in its infancy, but it is hoped that the material will prove stimulating for more research at the basic level.

## II. The Objectives of Freeze-Drying

For many years the process of freeze-drying, lyophilization or cryodesiccation has been employed primarily with the purpose of preserving living materials, including microorganisms, for extended periods of time; this is possible because, upon dehydration, substances no longer change as a consequence of the usual turnover of metabolic reactions characteristic of the living condition (REY, 1959); cells enter then into what could be refered to, as a state of cryptobiosis (HINTON, 1966).

Many excellent books and reviews have been published that discuss with great detail the technical aspects of lyophilization (REY, 1960, 1964). With particular reference to microorganisms, especially bacteria, HECKLY (1961) has written a very careful review on the many variables participating in the process and the book edited by MERYMAN (1966), contains several articles on various aspects of freeze-drying of microorganisms. We nevertheless consider it pertinent to remind the non-specialist that lyophilization involves the rapid freezing of the material and the subsequent sublimation of solidified water *in vacuo*. There are a large number of different techniques, mainly due to the large number of variables participating in the process, such as velocities of freezing, lowest temperature attained, type of protective colloids, suspending menstruum, storage temperature, storage atmosphere, residual moisture, reactivation conditions, etc., but, if the procedures have been correctly handled, the final result is that the water content of the samples under treatment will be only a fraction of a percent of the original preparation. As opposed to the destructive action of drying from the liquid state, the removal of water from material previously frozen under proper conditions, allows the maintenance of three major characteristics of the specimen: morphology, solubility and chemical integrity (MERYMAN, 1960).

## III. Effects of Freeze-Drying on Genetic Stability

Permanence of characteristics in microbial strains reflects genetic stability, manifested by constancy of biochemical, antigenic and physiological determinants, conversely, variation in these characteristics indicates genetic instability (LINCOLN, 1960). Although in theory an ideal system for preservation of cell characteristics, in practice, lyophilization has been shown to introduce permanent changes in the characteristics of some members of the treated populations; the appearance of these variants is explained most feasibly as due to modifications of the genotype, transmitted to the progeny and/or to selection of preexisting mutants present in the parental populations. Selection is due to genetically controlled differential resistance of some of the bacteria to the killing effect of the freeze-drying process. Such mutants will tend to increase in relative amounts in the population after reactivation from the lyophilized state. Selection by lyophilization has been shown to occur in mixtures of different microorganisms by GREAVES (1960), LEACH and SCOTT (1949) and GROSSBARD and HALL (1963), and more specifically between wild type and *E. coli* mutants by dehydration (WEBB and TAI, 1968). It is also possible that cryodesiccation may induce phenotypic effects by alteration of structures or molecules other than genetic material, but these will not be inherited by their progeny and the population, most of which is affected in this instance, will regain their original phenotype after a few generations.

It is known that the frequency of spontaneous appearance of mutants is low, but under treatment with a variety of physical or chemical agents, the proportions of mutants can be increased (ZAMENHOF, 1963); therefore, a quantitative way of thinking about possible mutagenic effects of freeze-drying on bacterial populations is necessary which will permit critical experiments discriminating mutant clones among the frozen-dried and reactivated population (DAVIS et al., 1968). But even though a determination of the exact frequency of mutants before and after treatment can be made with relative ease (BRAUN, 1965), it is striking to note, on reviewing the extensive literature on the problem of conservation of characteristics in lyophilized cultures, the predominance of reports in which it has been assumed, *a priori*, that again or loss of a character should take place in all, or most of the individuals in a treated population. The situation is reminiscent of the Lamarckism that used to enter into bacteriologists' way of thinking some years ago when bacterial variation was thought of in terms of adaptations by an entire population and not in terms of clones derived from mutants (HAYES, 1968). This state of affairs is unfortunate because it has frequently produced a failure to test for the possible occurrence of genetic changes induced by preservation methods, and it has led to a failure of applying proper quantitative methodology to the determination of the proportions of mutants. There are, nevertheless, a number of publications in which awareness of this point has indicated that direct genetic effects can occur as a result of lyophilization, and many other results can be interpreted in terms of increases of genetic variants after lyophilization of microbial cultures, even when the study was not supported by strict genetic analyses. Only in a few exceptional cases, appropriate determinations of selective killing have been made and, therefore, it is often uncertain whether selection only is the responsible mechanism in cases where the preservation process led to low survival values. It is difficult, then, in the case of most communications to assess whether the unusual increment in variants was due exclusively to a direct action of lyophilization on the genetic material of the microorganisms or was due to selective phenomena. In the light of all available data it seems very probable that both types of effects are produced by cryodesiccation. Let us examine the literature most relevant to the subject, calling attention to the fact that recently some material closely related to this review was published (NEI, 1969) and will not be covered by the present review.

### a) Colonial Variations

The possibility that preservation by lyophilization could modify the genotype of treated microorganisms was raised seriously for the first time by BRAUN (1950), who was able to identify changes in *Brucella* cultures after

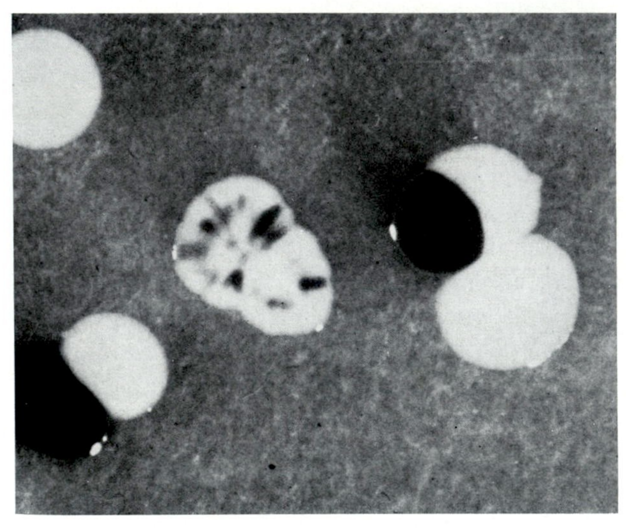

Fig. 1. Colonies obtained from cryodessiccated and subcultured *S. marcescens* samples after 48 hours of incubation at 30° C on Pennassay Seed Agar (Difco). Prodigiosinless colonies were found in proportions as high as 50%, sectored colonies in about 10%. Approximate magnification ×4. (From SERVIN-MASSIEU and CRUZ-CAMARILLO, 1969)

treatment. These changes consisted in the appearance of about 1 % non-smooth colonial types after freeze-drying, while in untreated bacterial populations only colonies of the S type could be identified. The author attributed the effects to possible selective survival of mutants having higher resistance to the freeze-drying process itself and to a direct increase in the number of mutants induced by the treatment (BRAUN, 1965). On the other hand, FLOSDORF and KIMBALL (1940) reported on the maintenance of the smooth form in *Bordetella pertussis* cultures preserved by lyophilization.

Another type of colonial variation attributed to freeze-drying has been reported for fungi; this is the case of increased frequency of mutations and reversions in giant colonies of baker's yeast. These mutational changes were manifested by sectoring of colonies, a phenomenom that was not apparent in non-lyophilized cultures (SUBRAMANIAM and PRAHALADA-RAO, 1951). These authors concluded that the method of preservation did produce genetic alterations primarily due to the cold "shock" involved in the freezing of the material during lyophilization, since they had been able to identify similar mutations in yeast maintained at "cold room" temperatures (SUBRAMANIAM et al., 1948). However, many cases of so called low-temperature "shock" actually represent slow velocity freezing which is known to cause intense intracellular dehydration (MAZUR, 1966); see IV a. A similar type of colonial variation has been observed after freeze-drying of *Staphylococcus aureus* (SERVIN-MASSIEU, 1961). In this system the dried bacteria showed, after rehydration and one subculture, an unusually high number (10%) of sectored and pigmentless colonies, in comparison to untreated cultures that produce uniformly pigmented colonies due to their capacity to synthesize carotenoid pigments (SUZUE and TANAKA, 1959). This sector formation effect in *S. aureus* can be most easily interpreted as the result of mutagenic effects induced by the cryodesiccation process (UMBREIT, 1962). Similar effects have been noted following lyophilization of *Serratia marcescens*, which commonly synthesizes a conspicuous red pigment allowing extensive genetic studies of color inheritance in bacteria (BUNTING, 1946); sectored colonies with pigmentless zones like the ones shown in Fig. 1, appear in high numbers in lyophilized, rehydrated and subcultured bacteria in addition to stable and unstable pigmentless variants (SERVIN-MASSIEU and CRUZ-CAMARILLO, 1969). The *Serratia* colonies with sectors have been shown to contain elevated numbers of bacteria with unstable pigment genes that cause the bacteria to sector again when resuspended and plated (Table 1). Other characteristics were also affected in these bacteria, such as respiratory mechanisms and the capacity to synthesize an inducible protease. The persistence of pigmentation genes in an unstabilized condition for many generations after freeze-drying and rehydration, closely resembles

Table 1. *Analysis of several colonies obtained by plating lyophilized Serratia marcescens*[a]. (From SERVIN-MASSIEU and CRUZ, 1969)

| Colony no. | Sector[b] | Percent of daughter colonies which were | | |
|---|---|---|---|---|
| | | pigmented | pigmentless | sectored |
| 1 | 15/16 | 2.3 | 97.0 | 0.7 |
| 2 | 15/16 | 0.4 | 96.0 | 3.6 |
| 3 | 7/8 | 2.0 | 93.5 | 4.5 |
| 4 | 2/5 | 67.0 | 30.9 | 2.1 |
| 5 | 1/4 | 87.6 | 11.8 | 0.6 |
| 6 | 1/16 | 89.4 | 9.5 | 1.1 |
| 7 | 1/16 | 93.0 | 6.2 | 0.8 |
| 8 | undetermined | 5.3 | 93.8 | 0.9 |
| 9 | stable pigmentless | 0 | 100.0 | 0 |
| 10 | wild type pigmented | 100.0 | 0 | 0 |

[a] Suspensions of excised colonies were plated at suitable dilutions on Pennassay Seed Agar (Difco) and incubated for 48 hs at 30° C.

[b] Approximate dimensions of pigmentless sector, relative to colony size.

sector formation produced in *S. marcescens* colonies after treatment with ultraviolet light, a known mutagen (KAPLAN, 1952). In the study of this type of variation, it is necessary to design appropiate controls that one can distinguish sector formation from certain artifacts (WITKIN, 1951).

The phenomenom of sector formation is considered a rare event, reflecting a genotypic variation due to an alteration of deoxyribonucleic acid and segregation during colony formation (LAMANNA and MALLETTE, 1965). When microorganisms are treated with well-known mutagenic agents, such as ultraviolet light, heat or X-rays (ZAMENHOF, 1961; BRAUN, 1965; WITKIN, 1951; NEWCOMBE, 1953), the phenomenom becomes more frequent, just as it seems to be the case following freeze-drying.

Sector formation in microbial colonies from lyophilized stock cultures also has an applied interest, especially in industrial microbiology where the appearance of such colonies is considered a reliable index of genetic instability in the testing of fermentation processes (LINCOLN, 1960).

## b) Antigenic Variations

Many strains employed for vaccination are frequently kept in a cryodesiccated condition and this has created a great deal of interest in searching for potential alterations in the antigenic determinants of the microbial cultures preserved in this way. For instance, it has been reported that lyophilization of *Salmonella paratyphi* A and *Salmonella paratyphi* B induces changes in antigenicity that remain manifest after several subcultures, especially for antigen H (LAMBIN et al., 1958). The same author has reported that six successive lyophilizations of subcultures, at one week intervals, did not modify significantly the formation of antigen O, but the capability to form antigen H was very much affected in these cultures. In the case of *Salmonella typhi* it was also observed that there was a loss in antigen Vi, however, effect on protective antibody formation was observed.

VELU et al. (1942) reported that after lyophilization *Pseudomonas mallei* alters its growth characteristics, such as rate of multiplication and morphology, and that animals injected with dried and subcultured bacteria were not immunized. The effect seemed to be a stable variation, lasting for many generations.

A report by SHARPE and WHEATER (1955), contained the results of a study on the conservation of physiological and serological characteristics in several lyophilized *Lactobacilli*. Of the 41 strains that were dried, rehydrated and subcultured daily for four days before testing with homologous sera, half the strains gave similar agglutination reactions before and after treatment, eleven strains gave a titer either one tube higher or one tube lower than the corresponding nonlyophilized strains, one strain gave a four fold lower titer after drying, one gave originally no reaction and after drying gave a high titer, and two strains gave autoagglutination reactions after freeze-drying (Table 2). It is difficult to assess the significance of these results since agglutination reactions are quite variable (KABAT and MAYER, 1963), but at least in two strains, the

Table 2. *Differences in agglutination reactions of lactobacilli after freeze-drying.* (From SHARPE and WHEATER, 1955)

| Strain | Agglutination titres against homologous type sera | |
|---|---|---|
| | before drying | after drying |
| *L. acidophilus* BF 4 | 640 | 1,280 |
| *L. bulgaricus* Y 48 | 80 | 160 |
| *L. leichmannii* LE 6 | 2,560 | 5,120 |
| *L. lactis* AH 7 | 320 | 640 |
| *L. buchneri* BC 1 | 320 | 640 |
| *Lactobacillus* sp. AH 4 | 320 | 640 |
| *Lactobacillus* sp. J 1 | 640 | 1,280 |
| *L. bulgaricus* B 2 | 1,280 | 640 |
| *L. casei* C 28 | 80 | 40 |
| *L. casei-helveticus* O 9 | 5,180 | 2,560 |
| *L. fermenti* F 1 | 2,560 | 160 |
| | | very slight agglutination |
| *L. fermenti* AH 18 B | no reaction | 2,560 |
| *Lactobacillus* sp. RF 1 | 40 | auto-agglutination |
| *L. brevis* SL 15 | 20 | auto-agglutination |
| *L. brevis* X 2 | auto-agglutination | auto-agglutination |

21 other strains of *Lactobacilli* gave equal titers before and after lyophilization; additional 4 strains were not tested.

Desiccated tubes were opened after 6 months of storage, reactivated in tomato glucose broth and subcultured four times. Cells were washed, suspended, standardized and tested by tube agglutination concurrently with original unlyophilized strains.

differences obtained were of sufficient magnitude to suggest that some variation in antigenic determinants had taken place. Using a similar methodology, and two strains of *Paracolon* and *Salmonella*, JENNENS (1954) failed to observe changes in antigenicity after lyophilization.

In a study of BCG vaccine (VAN DEINSE, 1951) it was observed that the appearance of "allergy" in guinea pigs innoculated with dry vaccine lagged many days behind the one provoked following the innoculation of fresh BCG vaccine and employing an equivalent basis of bacterial mass. This phenomenom may have been due to some variation, probably phenotypic in nature, of antigenic determinants. BIRKHAUG (1951) also observed a slower conversion with dried BCG vaccine than with freshly prepared bacteria.

In contrast STEELE and ROSS (1963) compared 100 strains, representative of fifteen bacterial genera, in regard to survival after freeze-drying and claimed to have found no changes in cultured, biochemical, serological or pathological characteristics, but no data were offered supporting this conclusion. Similarly, STILLMAN (1941) and SWIFT (1937) reported that several bacteria preserved by lyophilization did not suffer variations in their antigenicity, but it should be noted that their method of drying has been questioned (HECKLY, 1961).

### c) Virulence Variations

Data on this type of variation are scanty, but nevertheless there are a few communications to be mentioned. One of the most critical, deserving detailed examination because of the careful quantitative method followed, is that of PRIESTLEY (1952), who reported the effects of freeze-drying on viability and virulence characteristics of bovine pleuropneumonia organisms employed in cattle vaccination. Preliminary work established optimal conditions for centrifugal freeze-drying of the organisms and the testing of the attenuated strains utilized. They were injected into cattle after many weekly passages in serum broth medium and were found to be avirulent. They were then cryodesiccated and kept in an ice chest. Individual samples were used for vaccine production using not more than 10 subcultures for any given line. Most of the strains dried and employed did not show untoward results, but one of the strains, from which 114,000 doses were prepared and distributed, produced "thousands of diseased animals and hundreds of deaths in vaccinated cattle". A closer examination (Table 3) showed that freeze-drying of attenuated

Table 3. *The effect of freeze-drying on the virulence of attenuated cultures of bovine pleuropneumonia organism.* (From PRIESTLEY, 1952)

| Before drying | | | After drying | | |
|---|---|---|---|---|---|
| strain | generation number | results in cattle | strain | number days dried | results in cattle |
| F | 43 | 0/3 | F46 | 1,076 | 1/3 |
| F48 | 14 | 0/3 | F48 | 219 | 3/4 |
| 139 | 45 | 0/3 | F62 | 101 | 1/2 |
| | 49 | 0/3 | 139/51 | 18 | 2/3 |

Numerator: number of cattle dying and/or showing swellings.
Denominator: number cattle inoculated.

Before drying, avirulent strains were tested by inoculation into cattle after 43, 14, 45 and 49 weekly subcultures in serum broth. Cultures were then dried and stored in an ice chest. Tubes were opened and reactivated in serum broth; further subculturings were made to test for purity and production. The final vaccine was a four-day old growth of tested strains.

cultures increased their virulence properties sharply. PRIESTLEY was able to identify this important effect because he gathered information from a large number of individual observations (vaccinations) involving single lyophilized strains; this was equivalent to the testing of individual clones, much like a fluctuation test for spontaneous mutations in bacteria (LURIA and DELBRÜCK, 1943).

STEIN (1949), using *Pasteurella bubaliseptica*, also found evidence suggestive of an enhacement of virulence in cultures preserved by cryodesiccation, compared to strains maintained on laboratory culture media. In contrast, STAMP (1947) and STILLMAN (1941) published evidence indicating that virulence characteristics remained unaltered in lyophilized strains, but their technique could not differentiate quantitative changes among the treated bacteria.

In the field of soil microbiology, APPLEMAN and SEARS (1946) found that cultures of *Rhizobium leguminosarum* retain their nitrogen-fixing capabilities after lyophilization. HECKLY (1957), has also found that freeze-drying of *P. pseudomallei* does not alter virulence characteristics, in contrast to lyophilized *Pasteurella pestis* cultures which, when tested immediately after reconstitution, gave much lower virulence characteristics. However, after one subculture these strains recovered their normal virulence (HECKLY et al., 1958), indicating a phenotypic effect.

Discrepancy in results regarding the maintenance of virulence after lyophilization results probably from differences in techniques of drying and, most important, from the method followed for the evaluation of virulence. It would be interesting to learn more about this type of variation in other systems employing attenuated, lyophilized microorganisms for vaccination.

### d) Variations in Resistance

Techniques for the evaluation of this type of variation, although relatively simple, have not been very frequently used for the investigation of changes after cryodesiccation. One of these studies has analyzed the frequency of streptomycin-resistant mutants in cultures of *Staphylococcus aureus* (SERVIN-MASSIEU, 1967). In tests with several antibiotic concentrations it was found that a lyophilized culture (A.T.C.C., 1964) showed an approximately ten-fold increase in the frequency of resistant mutants and this effect was not due to a selective killing of the parental antibiotic-sensitive population (Fig. 2). The procedure for testing this involved direct isolation of several of the resistant mutants and a determination of their viability ratios before and after freeze-drying compared to ratios obtained with samples from the wild type sensitive strain. No significant differences in survival due to the killing action of the lyophilization process were found.

It also has been possible to demonstrate that lyophilization induces significant changes in the bacteriophage typing pattern of *S. aureus* (Fig. 3.). In this study, SERVIN-MASSIEU et al. (1968), prepared parallel cultures in liquid media from lyophilized and from unlyophilized samples, which were then plated so as to obtain isolated colonies. 100 colonies were picked from each parallel culture and individually subcultured in a small amount of broth. Each subculture was then spread on the surface of separate plates and the lytic patterns were determined after the addition of drops from phage suspensions. It was possible to determine changes in the lytic pattern of *S. aureus* type strains 81, 3a and 7, which have narrow, intermediate and wide phage patterns respectively. Although the results need further confirmation they suggest that aside from a genetic instabilization there may be some alterations in immunity due to prophage particles frequently carried by most *S. aureus* strains (ADAMS, 1958; ELEK, 1962).

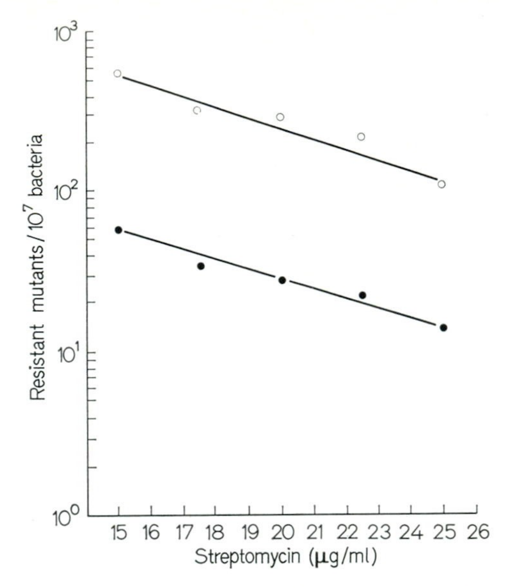


Fig. 2. Comparative survival curves of lyophilized (o — o) and non-lyophilized (● — ●) *S. aureus* T-81-CDC cultures. Parallel cultures, in liquid medium, were washed, adjusted to appropiate concentrations and plated on media containing different concentrations of the antibiotic. (From SERVIN-MASSIEU, 1967)

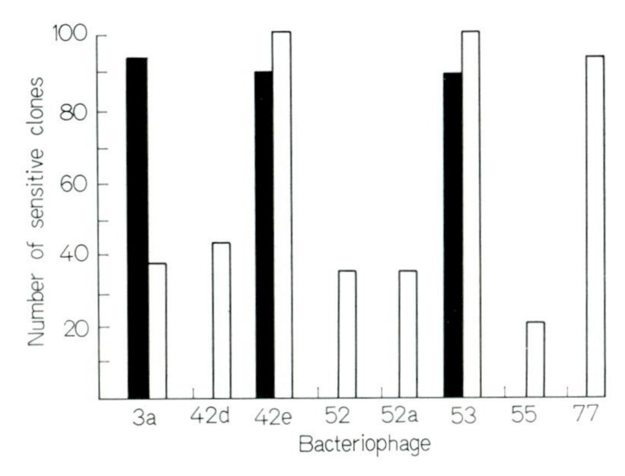


Fig. 3. Lytic pattern of isolated clones of *S. aureus*, type strain 7, unlyophilized (■) and after lyophilization (□). Sensitivity to phages 6, 7, 47, 54, 81, 83a, d, and 77ad is not indicated, but was also tested, giving 100% sensitivity before and after lyophilization. (From SERVIN-MASSIEU et al., 1968)

## e) Physiological Variations

This type of variation has received much attention, mostly because the microbiological industry is aware that its most valuable working capital is in its collection of stock cultures of well defined microbial strains, which must be constant in their ability to produce useful compounds in high yields (WIKEN, 1963), but, surely, are not always so, a fact that is largely unpublished for confidential reasons (SIMMONS, 1963).

There are several reports regarding genetic variation in growth requirements after lyophilization. BRAENDLE (1963), for example, found increased reversion frequencies in *Penicillum chrysogenum* and *Nocardia* auxotrophs (Tables 4

Table 4. *Genetic stability of lyophilized Penicillium auxotrophs.* (From BRAENDLE, 1963)

| Culture | Requirement | Before lyophilization | | After lyophilization | | Killed (%) |
|---|---|---|---|---|---|---|
| | | cells tested | proto-trophs | cells tested | proto-trophs | |
| 1C2-3 | $choline_1$ | $2.8 \times 10^5$ | 0 | $5.1 \times 10^5$ | 0 | 89 |
| 2C1-1 | $choline_2$ | $3.8 \times 10^6$ | 0 | $4.9 \times 10^6$ | 1 | 35 |
| 2C1-2 | arginine | $1.3 \times 10^6$ | 1 | $1.7 \times 10^6$ | 1 | 32 |
| 1C2-1 | isoleucine | $4.7 \times 10^6$ | 1 | $9.8 \times 10^5$ | 1 | 70 |
| 2C1-3 | nicotinamide | $4.4 \times 10^6$ | 8 | $3.1 \times 10^6$ | 9 | 65 |
| 2C1-3 | nicotinamide | $2.3 \times 10^7$ | 9 | $1.4 \times 10^7$ | 9 | 94 |

Table 5. *Genetic stability of lyophilized Nocardia auxotrophs.* (From BRAENDLE, 1963)

| Culture | Requirement | Before lyophilization | | After lyophilization | | Killed (%) |
|---|---|---|---|---|---|---|
| | | units tested | proto-trophs | units tested | proto-trophs | |
| 1L2-2 | tryptophan | $6.5 \times 10^8$ | 0 | $8.2 \times 10^8$ | 0 | 37 |
| 1L2-6 | arginine | $7.0 \times 10^8$ | 0 | $8.8 \times 10^8$ | 0 | 32 |
| 1L4 | arginine | $2.8 \times 10^7$ | 0 | $2.8 \times 10^7$ | 2 | 52 |
| | | $3.3 \times 10^8$ | 0 | $1.6 \times 10^8$ | 6 | 76 |
| | | $4.3 \times 10^9$ [a] | 0 | $2.6 \times 10^9$ | 0 | 25 |
| | | $1.8 \times 10^7$ | 2 | $9.8 \times 10^6$ | 23 | 24 |
| | | $4.0 \times 10^7$ | 11 | $2.2 \times 10^7$ | 23 | 32 |

[a] The last three cultures were grown for 4, 14 and 20 days before freeze-drying.
No evidence was found for population changes occurring during the storage of lyophilized cultures for periods of up to three months. Viability losses during this periods were the same for prototrophs and for auxotrophs.

and 5). This effect was particularly evident if the cultures were "aged" before processing. No evidence was found that could suggest selective killing of either the prototrophs or the auxotrophs.

VOLZ and GORTNER (1948) have shown that the lyophilization of *Lactobacillus casei* provoked a diminished capacity for lactic acid production in the presence of riboflavin, a change that did not disappear after several subcultures. SHARPE and WHEATER (1955) found a low frequency of variation for several physiological markers of *Lactobacilli*, including fermentation characteristics; the changes were identified qualitatively and the authors attributed them to a selective killing effect of the freeze-drying process. ROGERS (1914) did not find any alterations in the production of lactic acid by *L. bulgaricus* subjected to freeze-drying.

WASSERMAN and HOPKINS (1958) observed alterations in enzymes of lyophilized *Serratia marcescens* involved in the oxidation of 2-oxo-gluconic acid, but the data seemed to suggest a transient phenotypic effect from which the bacteria recovered after some generations. SERVIN-MASSIEU and CRUZ-CAMARILLO (1969) have been able to identify permanent changes in *S. marcescens* cultures after lyophilization; in addition to inabilities in the capacity to synthesize the characteristic pigment of this bacterium, prodigiosin, some pigmentless mutants showed a diminished capacity to synthesize an inducible protease and also displayed alterations in respiratory patterns when compared with wild type pigmented bacteria. Changes in respiratory patterns have also been found for other bacteria and yeasts after freeze-drying (TOKIO-NEI, 1960), but no mention is made about the permanence of these alterations. The author observed an increased effect with increasing cooling rate in the range from 3° C/sec to 1° C/min. The author also observed by electron microscopy that virions of tobacco mosaic virus were broken into several fragments by the freeze-drying process, and that their infectivity was significantly lowered.

In a study covering a ten year period, HARRISON and PELCZAR (1963) studied viability and physiological characteristics of twelve microbial species before and after lyophilization. No change in characters was noted in most of the strains, except in two *Bacteroides* strains, in which losses in the ability to ferment polyhidric alcohols, di- and tri-saccharides were noted; one of these strains, in addition, had lost the ability to ferment glycerol. Although the report contains no quantitative data regarding the survival of the two strains to cryodesiccation, indirect evidence allowed the authors to attribute the effects to a selective survival of variants of different genotype in which the fermentative properties are linked to a higher resistance to the freeze-drying process.

JANUSZEWICZ (1957) also observed that cultures of *Leuconostoc mesenteroides* underwent changes in growth requirements after the strains were lyophilized and, most important, noted an increased capacity to synthesize dextran compared with non-desiccated cultures. The author explained these variations as possibly due to selective phenomena, but the data do not exclude the possibility of a direct effect of dehydration on genotype. Similar conclusions have been reached by ATKIN et al. (1949) in the case of low survival of brewer's yeast after freeze-drying. Treated samples, after rehydration, were plated to obtain isolated colonies; these were separated subcultured and tested for vitamin requirements. The data obtained showed a surprisingly high number of "gain" or "loss" mutants for biotin, pantothenate, inositol, $B_1$ and $B_6$.

Several authors have specifically commented that no variations in fermentative characteristics of lyophilized microbial strains can be identified, but unfortunately no supporting data are usually offered (MARTIN, 1963). On the other hand, detailed experiments on genetic variation in *Saccharomyces pastorianus* after freeze-drying were performed by BRADLEY (1963) who found no effects on enzyme systems involved in sugar utilization at constitutive and inducible levels.

## IV. On the Origin of Changes after Lyophilization

There are four principal operations involved in freeze-drying which, potentially, could be responsible for the genesis of the observed variations, either at the level of selective killing or at the level of direct action on the hereditary material. These operations are: a) freezing of the material, b) dehydration, c) storage of the dried biological material and d) rehydration (MERYMAN, 1966). The first two operations have received principal attention in regard to the problem of variation and their effects have been followed separately by studying, respectively, freezing and drying; one could take these studies to serve as model systems for equivalent stages in the overall freeze-drying process.

### a) Effects of Freezing on Microbial Variation

Factors responsible for cellular injury in freezing of microorganisms and cells have been reviewed recently by MAZUR (1966) and NEI (1969). Regarding changes in the characteristics of cells, there are some reports suggesting that freezing may produce them. For instance, POSTGATE and HUNTER (1963) were able to identify four auxotrophic mutants among 6,200 colonies screened by replica plating, of a culture of *Aerobacter aerogenes* that had been exposed to $-196°$ C. In addition they observed numerous cells that were "metabolically injured". SUBRAMANIAM et al. (1951) have also reported that they were able to isolate sectored colonies after subjecting yeast cultures to "cold shocks". STRAKA and STOKES (1959) obtained evidence of metabolic injuries, of unreported permanence, by exposing bacteria to freezing at $-78°$ C and switching to conservation at $-7°$ C, $-18°$ C and $-29°$ C. Similar results were obtained by ARPAI (1962).

Aside from the development of ice crystals under conditions that have a close relationship to killing of the cells (MAZUR, 1966), the only known immediate result of freezing is dehydration (MERYMAN, 1966) but nothing is known about the possible contribution of this concurrent desiccation on changes in microbial properties.

It is improbable that freezing alone has little direct effect on genetic alterations. This can be inferred from the stability of the characteristics of a large number of different strains of *Hemophilus influenzae, Bacillus subtilis,* and *Bacillus licheniformis* preserved routinely at $-70°$ C and $-40°$ C in genetic transformation studies. In these systems transformable cells, at a definite stage of their growth cycle, are frozen rapidly and kept at the specified temperature for several weeks. No significant increases in variant proportions have been reported as a consequence of this procedure.

Effects of freezing on deoxyribonucleic acid have also been investigated. SHIKAMA (1965), was unable to find alterations in the structure of the macromolecule on the basis of spectrophotometric determinations. Similar conclusions were obtained by CABRERA-JUAREZ and OLGUIN (1968) who investigated the possible induction of new mutations to antibiotic resistance by freezing of transforming DNA from *H. influenzae*. Employing one cooling velocity, the authors froze samples of transforming DNA down to −70° C, and maintained them at that temperature for various periods of time after which samples were thawed and tested in genetic transformation assays with antibiotic-sensitive strains of *H. influenzae* (GOODGAL and HERRIOTT, 1961). No new genetic markers, of the type screened, were observed in frozen-undenatured DNA or in denatured-frozen-reannealed samples (Tables 6 and 7).

Table 6. *Effect of successive freezing and thawing of Hemophilus influenzae $Sm_{250}$ DNA* [a]. (From CABRERA-JUAREZ and OLGUIN, 1968)

| Freezing and thawing cycles [b] | Intrinsic transforming activity [c] | New markers [d] |
|---|---|---|
| Unfrozen DNA | 99 | 52 |
| 1 | 100 | 50 |
| 2 | 99 | 43 |
| 3 | 111 | 50 |
| 4 | 82 | 45 |
| 5 | 92 | 68 |
| Control without DNA | 0 | 40 |

[a] Genetic marker $Sm_{250}$ confers resistance to at least 250 μg/ml of streptomycin.
[b] Samples stored frozen for 30 minutes.
[c] $10^4$ transformations per μg of DNA.
[d] $10^2$ stable mutants to 30 μg of kanamycin per ml of transforming mixture.

Table 7. *Effect of freezing and thawing on Hemophilus influenzae $C_{25}$ DNA denatured by heat* [a]. (From CABRERA-JUAREZ and OLGUIN, 1968)

| Sample | Intrinsic transforming activity [b] | New markers [c] |
|---|---|---|
| A. Denatured and unfrozen | 93 | 152 |
| B. Denatured frozen and thawed | 99 | 120 |
| Annealed from A. | 2,180 | 162 |
| Annealed from B. | 2,250 | 198 |
| Control without DNA | 0 | 143 |

[a] Genetic marker $C_{25}$ confers resistance to 25 mcg/ml of cathomycin.
[b] $10^2$ transformations per microgram of DNA.
[c] $10^2$ resistant mutants to 40 μg of streptomycin per ml of transforming mixture.

Certain types of injury due to freezing have been reported, but these appear to be principally phenotypic effects from which cells recover after some time if environmental conditions are favorable. For instance SOUZU and ARAKI (1962) observed evidence of injury to nucleic acid metabolism after freezing yeast, and WASSERMAN and HOPKINS (1958) have found destruction and/or inactivation of *S. marcescens* enzymes involved in the oxidation of 2-oxo-gluconic acid after freeze-drying. Some of these deleterious effects probably are due to alterations in some of the molecules important for normal cell functioning. It is known that freezing, can cause alterations in proteins (LEIBO and JONES, 1964; LEVITT, 1962, 1966; LEA and HANNAN, 1950), phospholipids and lipoproteins (LOVELOCK, 1954, 1955), polyphosphates (SOUZU, 1967a), nucleotides (SAITO and ARAI, 1957), enzymatic proteins (TAPPEL, 1966; HANAFUSA, 1969) and of cell integrity (NEI, 1960; SOUZU, 1967b; HANSEN and NOSSAL, 1955).

## b) Effects of Drying on Microbial Variation

Freezing is a form of dehydration in which intracellular water is removed from the cell as effectively as by drying. However, bound water, which comprises 5 to 10% of the cells' total, remains attached to the biological material under these conditions. When more drastic methods of dehydration are carried out, like cryodesiccation, bound water is eliminated, and this action has been considered injurious to biological materials in at least three ways: a) denaturation of proteins due to concentration of still unfrozen, undried portions of the sample, b) exposure of reactive proteins to deleterious reactions by removal of water molecules and c) recrystallization of salts or hydrates formed from eutectic solutions producing extensive mechanical injuries to structural elements. An additional form of injury by removal of bound water molecules is considered feasible, namely a permanent alteration in genetic material, producing permanent variations or mutations, which is transmitted to the progeny of cells that have suffered such injury to their DNA. Also, populations shifts in favor of one or another type of cell with different genotype may take place due to selective conditions, as has been pointed out before in Section III.

There are several reports devoted to optimal conditions of drying during lyophilization required for optimal survival of treated cultures, but few data have been presented on the maintenance of characteristics under different drying conditions (HECKLY, 1961). Nevertheless, there are some communications correlating desiccation of cultures and attendant variation in cell characteristics. MALTMAN et al. (1960), studied the effect of "room temperature" drying on some characteristics of *Staphylococcus aureus*. The authors deposited films of bacteria on the surface of glass tubes covered with agar, dried them for 48 hours, after which the tubes were stored for different periods of time and reconstituted with broth medium. In view of the simplicity of this method, it is surprising that survival values of approximately 11% were obtained after two days storage and 7% after 14 days. Drying was shown to have injured, non lethally, some of the cells that remained viable, leading to an alteration in several bacterial characteristics, including a longer lag phase of growth, decreased capacity to survive reconstitution in various fluids, and a decreased rate of coagulase production. The authors attributed these variations to damaged cell structures that caused a leakage of important cellular

Table 8. *Production of auxotrophic mutants by controlled desiccation of E. coli cultures.* (From Webb, 1967)

| Relative humidity | Experiment no. | Mutant type | | |
|---|---|---|---|---|
| | | amino-acid dependent | base dependent | vitamin dependent |
| 30% water | 1 | 114 | 18 | 6 |
| | 2 | 106 | 7 | 3 |
| | 3 | 92 | 16 | 4 |
| 40% water | 1 | 301 | 44 | 16 |
| | 2 | 322 | 36 | 11 |
| | 3 | 304 | 49 | 14 |
| 55% water | 1 | 255 | 22 | 11 |
| | 2 | 203 | 14 | 8 |
| | 3 | 268 | 36 | 4 |
| 75% water | 1 | 2 | 0 | 0 |
| | 2 | 3 | 1 | 0 |
| | 3 | 0 | 0 | 0 |
| 40% water + inositol | 1 | 4 | 1 | 2 |
| | 2 | 5 | 3 | 0 |
| 55% water + inositol | 1 | 8 | 0 | 3 |
| | 2 | 10 | 3 | 1 |
| Control | 1 | 1 | 0 | 0 |
| | 2 | 0 | 0 | 0 |
| | 3 | 3 | 1 | 0 |
| | 4 | 1 | 0 | 1 |

Cells grown for 48 hours in yeast extract broth, washed in water and desiccated for 60 minutes in nitrogen.

Mutant type numbers represent mutant colonies per $10^6$ viable cells.

components. In an extension of this investigation, the authors determined the effects of this drying method on the virulence of *Staphylococcus aureus* (Hinton et al., 1960), and found it also to be affected. Unfortunately as in their previous communication, no data were presented regarding the permanence of these alterations. Elimination of bound water from the neighborhood of protein molecules could explain alterations in enzymatic systems (Wasserman and Hopkins, 1958) including those studied in the tests of Maltman et al.

A different perspective to the problem of genetic stability after dehydration of cells was provided by the investigations of Webb (1965) who allowed a controlled desiccation of bacteria in aerosols of varying relative humidity. He showed that by lowering the relative humidity of the bacterial environment, several kinds of mutants were produced under conditions that excluded selective killing (Webb, 1967). He also investigated the effects of desiccation on *E. coli*

Table 9. *Production of auxotrophic mutants by controlled desiccation of E. coli cultures.* (From WEBB, 1967)

| Relative humidity | Experiment no. | Mutant type | | |
|---|---|---|---|---|
| | | amino-acid dependent | base dependent | vitamin dependent |
| 30% water | 1 | 124 | 12 | 10 |
| | 2 | 177 | 16 | 8 |
| | 3 | 111 | 5 | 3 |
| 40% water | 1 | 342 | 32 | 12 |
| | 2 | 316 | 31 | 16 |
| | 3 | 298 | 41 | 10 |
| 55% water | 1 | 408 | 66 | 22 |
| | 2 | 516 | 48 | 35 |
| | 3 | 564 | 57 | 48 |
| 75% water | 1 | 10 | 3 | 3 |
| | 2 | 4 | 0 | 8 |
| | 3 | 3 | 4 | 1 |
| 40% water +inositol | 1 | 42 | 10 | 4 |
| | 2 | 26 | 6 | 0 |
| | 3 | 18 | 4 | 3 |
| 55% water +inositol | 1 | 53 | 21 | 10 |
| | 2 | 72 | 18 | 12 |
| Control | 1 | 0 | 1 | 0 |
| | 2 | 0 | 0 | 2 |
| | 3 | 2 | 0 | 1 |
| | 4 | 0 | 2 | 0 |

Cells grown for 12 hours in yeast extract broth, washed in water and desiccated for 60 minutes in nitrogen.

Mutant type numbers represent mutant colonies per $10^6$ viable cells.

in the exponential and stationary phases of growth. For this purpose washed cell suspensions were atomized in a rotating drum in an atmosphere of nitrogen at 25° C. Relative humidity was controlled by spraying water free of oxygen into the drum and cells were treated for one hour at various relative humidity values. $10^7$ cells/ml were collected in a liquid impinger containing a solution of glucose and sodium chloride, were then cultivated in a yeast extract solution for 90 minutes at 37° C after which they were washed, starved and resuspended in a minimal salts medium. Auxotrophic mutants were then selected by the penicillin method, purified and tested in minimal media supplemented with pertinent additions. Control bacterial suspensions were treated the same way, except that the desiccation stage was omitted. Results presented in Tables 8 and 9 clearly indicate that desiccating the cells beyond a critical relative humidity value, significantly increases the frequency of various classes of mutants. It can also be observed that the frequency of mutants was higher when cells were treated at their exponential stage of growth. This effect was interpreted by the author as due to a higher sensitivity of the DNA to desiccation while in the process of replication. Inositol showed a marked protective effect against the mutagenic action of desiccation, presumably by producing a functional substitution of bound water eliminated by the drying of the macromolecule (WEBB and BHORJEE, 1968). Possible explanations offered were an irreversible attachment of protein to certain gene sites, or more probably, changes in DNA structure resulting from the dehydration process. The latter explanation is more plausible in view of additional spectrophotometric evidence showing a definite role of bound water in the maintenance of functional stability of the macromolecule (WEBB and DUMASIA, 1968). The extensive studies by this group also revealed that dehydration alters the characteristics of bacteriophage DNA in lysogenic bacteria in such a way that phage induction and DNA damage occurred below certain critical values of relative humidity (WEBB and DUMASIA, 1967a; WEBB and DUMASIA, 1967b). In tests on conjugation in *E. coli* these investigators learned that genetic recombination took place at higher efficiency after a short period of dehydration of male cells, suggesting that chromosome breakage may occur in certain, more susceptible, zones of the DNA (WEBB, 1968, 1969).

It can be understood, then, how systems involving a drastic drying stage, like freeze-drying, could affect the genotype directly, especially when the elimination of water is sufficient to remove bound water from the vicinity of deoxyribonucleic acid.

### c) Effects of Storage and Rehydration

These two variables of the lyophilization process have received little consideration in regard to the appearance of changes in preserved cultures, neither have several other aspects of storage and rehydration such as the atmosphere within dried vials (DAMJANOVIC et al., 1969), the temperature of storage (POPOVSKY, 1969), illumination conditions, moisture content (GHEORGHIU and STURDZA, 1969), type of container, temperature of the reconstitution fluid, its composition, volume, rate of rehumidification, etc. Some of these variables have been shown to produce deleterious conditions which, at least by selection, could eventually produce shifts in the proportion of genetic variants initially present (MAZUR, 1963).

The problem of cryoprotectants in lyophilization (O'CONNELL et al., 1968) has also important implications deserving further study, not only as substances that determine optimal survival and thereby checking potential selective effects, but also as agents protecting against probable mutagenic effects of the drying process. In this respect, the findings of WEBB and his group regarding inositol are most important, and would suggest that perhaps this substance, or others physiologically related to it, could be added to lyophilization menstrua as antimutagen (NOVICK, 1956) helping to preserve the genetic stability of the treated cultures.

## V. Effects of Freeze-Drying on Deoxyribonucleic Acid

Ever since the classical experiments of WATSON, CRICK and WILKINS on the structure of DNA, it has been recognized that changes in humidity can have a profound effect on the overall structure of the hereditary material. Two different configurations, A and B, were postulated on the basis of crystallographic evidence at two different humidity values (WATSON and CRICK, 1953; WILKINS, 1956). Since the macromolecule contains in its periphery negatively ionized phosphate groups, it can be considered as a highly charged and symmetric polyanion (COLE, 1967), able to bind water in a more or less structured form (JACOBSON, 1953; HEARST and VINOGRAD, 1961). Further study of the problem has confirmed that water surrounding the DNA molecule contributes to the maintenance of a stable conformation (GORDON and CURNUTTE, 1965) and that conditions involving removal of supporting water molecules, like dehydration, when applied to deoxyribonucleic acid, can bring about a collapse of the structure of the macromolecule (SUTHERLAND and TSUBOL, 1957; BRADBURY et al., 1961). This effect would eventually produce anomalous biological properties in DNA (SPITKOVSKII et al., 1960) like an interference with normal replication processes of this material, a possibility that was postulated many years ago (JACOBSON, 1953). Hydration and dehydration have long been suspected of having an important role in the control of cellular duplication and would reflect, still further, the susceptibility of DNA to abnormal hydration conditions in its environment (SERRA, 1955).

Table 10. *Effect of lyophilization on B. licheniformis arg⁺ DNA transforming capacity*[a]. (From SERVIN-MASSIEU et al., 1970)

| DNA concentration (μg/ml) | Transformants/$10^7$ arg$^-$ cells | |
|---|---|---|
| | unlyophilized | lyophilized |
| 1.0 | $0.7 \times 10^3$ | $0.9 \times 10^2$ |
| 11.0 | $1.2 \times 10^3$ | $1.8 \times 10^2$ |
| 54.0 | $7.5 \times 10^3$ | $3.1 \times 10^2$ |
| 108.0 | $40.0 \times 10^3$ | $5.6 \times 10^2$ |

[a] *B. licheniformis* wild type 9945-A was employed as source of DNA. Competent cells were prepared from a mutant M-18, requiring arginine following the procedure of Gwinn and Thorne (1964). DNA was lyophilized while disolved in 2 M saline and rehydrated in distilled water; this solvent has been shown to confer maximum stability to the macromolecule (SPIZIZEN, 1958) and protection to cells against freezing (MAZUR, 1963).

Thus it would not be very surprising, to expect that cryodesiccation may alter the properties of DNA *in vivo* and *in vitro*, and even though there is still no direct evidence for an induction of mutations by desiccation of purified transforming DNA, data at hand indicate that the macromolecule is profoundly altered in some of its biophysical characteristics by freeze-drying. For instance,

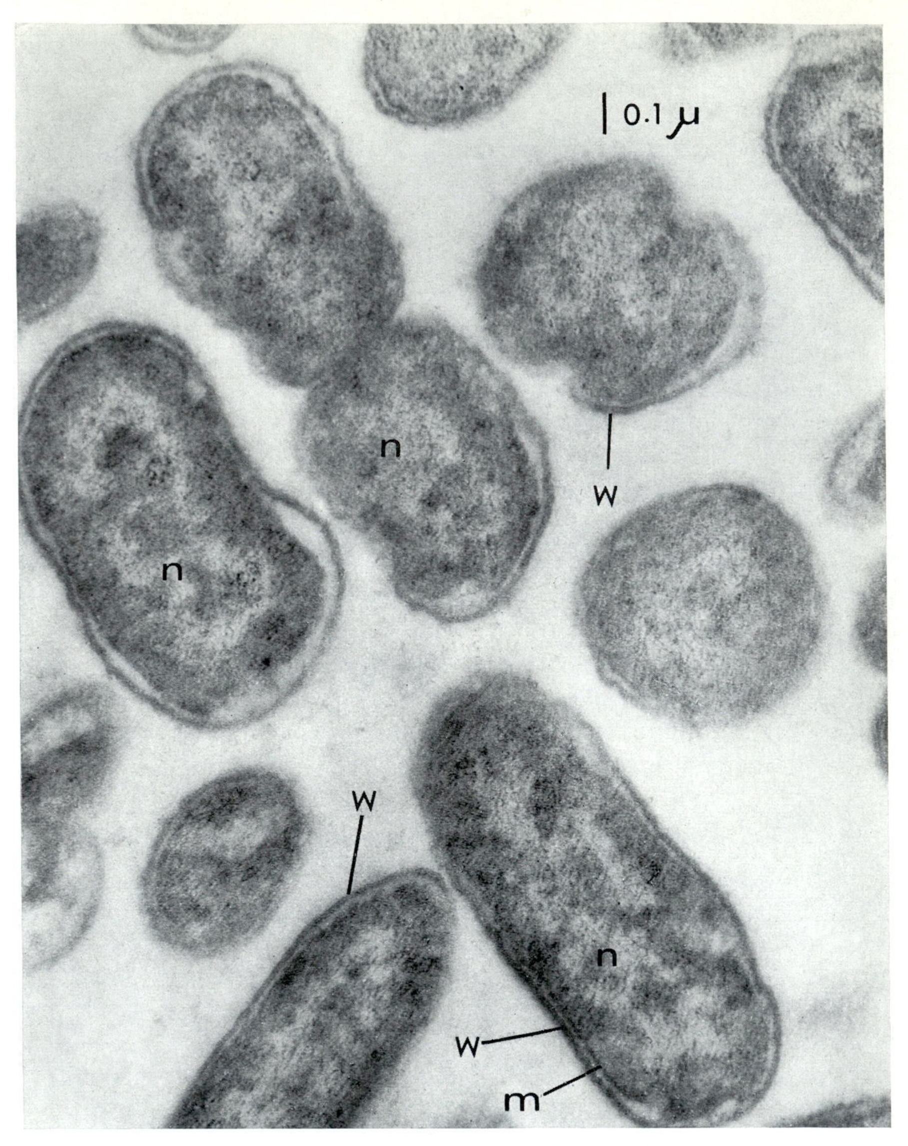


Fig. 4. Electron micrographs of ultrathin sections of unlyophilized *Serratia marcescens* cells. Nuclear structure is discernible from the rest of the cytoplasm, *w* cell wall, *m* cytoplasmic membrane, *n* nuclear material. (From PURKAYASTHA et al., 1961 and WILLIAMS, unpublished)

AVERY et al. (1944) described some time ago that lyophilization of purified transforming DNA from pneumococcus produces a marked loss in transformation capacity accompanied by a loss in solubility; however, no quantitative data on this effect were presented. More recently, SERVIN-MASSIEU (1969) has shown that lyophilization of purified transforming DNA from *Bacillus licheniformis* produces a significant decrease in viscosity of the macromolecule and, parallel to this effect, the cryodesiccated samples manifested strong hypochromicity and an intense loss in transforming capacity (Table 10). These findings are compatible with the idea that upon dehydration, the DNA molecules alter their conformation and possibly adopt a more compact form in an aggregated state (BALDWIN, 1968; SPITKOVSKII et al., 1960).

The effects of lyophilization on DNA within intact cells can be deduced from observations by Williams and his collaborators (PURKAYASTHA et al., 1960) who noticed that lyophilization of *Serratia marcescens* cells before fixation for electron microscopy, yielded a better resolution of the cell wall

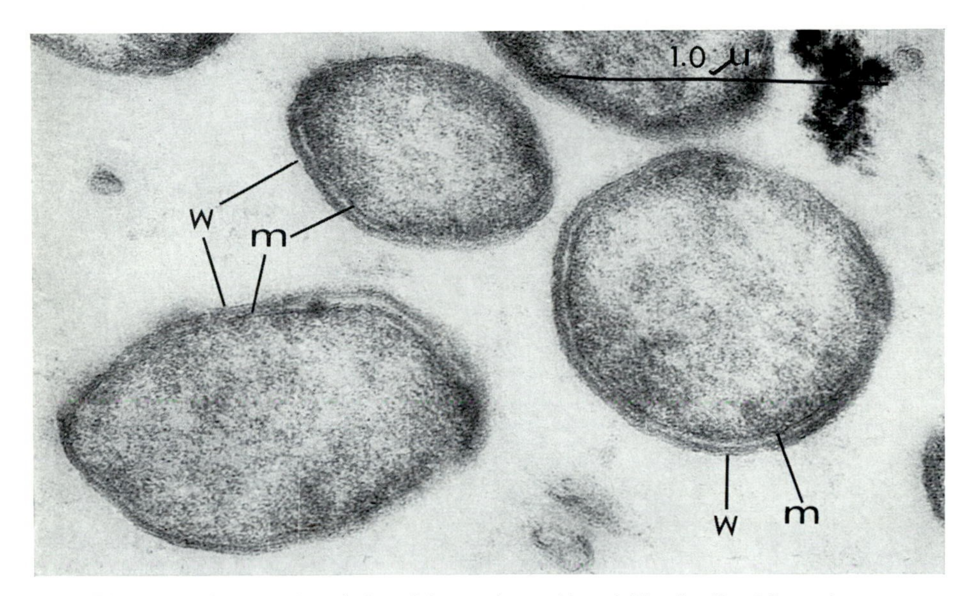


Fig. 5. Electron micrographs of ultrathin sections of lyophilized cells of *Serratia marcescens*. Discernible nuclear structure is lost and the cytoplasm is filled with granular material, *w* cell wall, *m* cytoplasmic membrane. Cells were lyophilized before fixation. (From PURKAYASTHA et al., 1961 and WILLIAMS, unpublished)

and of the cytoplasmic membrane, with an accompanying loss of nuclear structure and more prominent granulation (Figs. 4 and 5). How this effect in the nuclear material of the intact cell correlates with the effects obtained with isolated DNA remains to be studied, but most probably, both types of results are compatible with the idea of a change in the normal DNA configuration after removal of water.

## VI. Sporulation and Increased Mutagenesis

In view of the points discussed in Sections IV and V, it may be concluded that of the two crucial operations involved in the lyophilization process, freezing and drying, the former does not seem to be directly responsible for changes in the genotype of treated cells, whereas dehydration does seem to produce genetic changes by direct alterations of DNA; in addition it may cause selective phenomena. It would be of interest then to ask: what happens in regard to hereditary material in natural systems of dehydration, like sporulation? Would bacteria derived from germinating spores, for instance, show a comparatively higher frequency of mutation than their corresponding vegetative, hydrated forms?

Sporulation is known to be a natural process that some microorganisms, including bacteria, are able to carry out, in which, a portion of the cytoplasm and DNA is isolated by several coats of protective material from the rest of

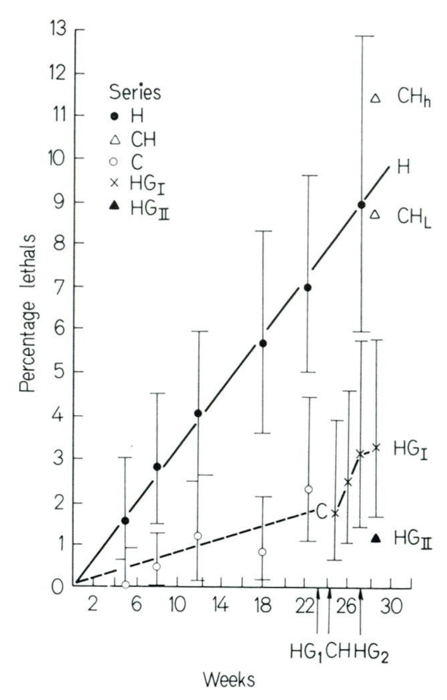


Fig. 6. Frequencies of recessive lethals in (*H*) conidia stored dry at 30° C, (*C*) conidia stored dry at 4° C, (*CH*) conidia stored dry at 30° C for 24 weeks and subsequently transferred to 4° C ($CH_l$ lower value, $CH_h$ higher value). ($HG_I$) and ($HG_{II}$) conidial samples from growth tubes that had been started with conidia from the (*H*) series at 23 and 27 weeks. Vertical lines: fiducial limits for a probability of 5%. (From AUERBACH, 1959)

the cell and kept in a relatively anhydrous form, with lowered metabolism (FRIEDMAN and HENRY, 1938; MURRELL and SCOTT, 1957; DAVIS et al., 1968). While the problem of absolute water content in spores has been controversial, the relative water content of these structures, although depending on the method of storage, is said to be much lower than in the corresponding vegetative form (ROSS and BILLING, 1957).

Data comparing mutant proportions in microbial spores and in vegetative cells are available. AUERBACH (1959) has identified increases in lethal mutations in stored conidia of *Neurospora* under dry conditions, and although the moisture content of her spore preparations was not strictly controlled, the progeny of dry spores stored at 30° C showed an increasing proportion of mutants with

Table 11. *Azide-resistant mutants in spores of Bacillus subtilis strain 23 stored in water*[a]. (From ZAMENHOF et al., 1968)

| Storage temperature | 3° C | | | 23° C | | |
|---|---|---|---|---|---|---|
| Length of storage months | 0 | 8 | 12 | 0 | 8 | 12 |
| Starting wild | | | | | | |
| Spore survival | 1 | $3.5 \times 10^{-1}$ | $9.4 \times 10^{-2}$ | 1 | $5.1 \times 10^{-2}$ | $3 \times 10^{-2}$ |
| Mutant frequency | $6 \times 10^{-6}$ | $2 \times 10^{-6}$ | $1.1 \times 10^{-5}$ | $6 \times 10^{-5}$ | $2 \times 10^{-6}$ | $2.5 \times 10^{-6}$ |
| Population mixture[b] | | | | | | |
| Spore survival | 1 | $7 \times 10^{-1}$ | $2.4 \times 10^{-1}$ | 1 | $1 \times 10^{-2}$ | $2.7 \times 10^{-3}$ |
| mutants (%) | 96 | 100 | 88 | 96 | 74 | 64 |

[a] Average of three determinations each. Washed spores of strain 23 in water ($10^8$/ml) were stored in sealed ampules at indicated temperatures for the times shown. Proportions of azide-resistant mutants before and after storage were determined as described previously by ZAMENHOF and EICHHORN (1967).

[b] Mixture of Az-resistant and Az-sensitive parents was stored as in [a] to determine probable selection; "reconstruction" experiment.

increasing time of storage. Samples stored at 4° C also manifested this phenomenom, but with lower intensity, due perhaps to moisture being absorbed during storage (Fig. 6). These results have been confirmed and extended in a different system by ZAMENHOF et al. (1968). These authors stored *Bacillus subtilis* spores at various temperatures and moisture conditions and evaluated for azide-resistant mutants. The results (Table 11) indicated, that no increases in azide-resistant mutants took place when spores were stored in water suspensions for periods of up to 12 months at 23° C or 3° C; however, when spores were subjected to desiccation (in glass bulbs over $P_2O_5$ under high vacuum) and stored for several time periods at 23° C, results differed completely (Table 12). Desiccation itself increased the mutant frequency 19-fold, and 7 months of storage brought the increase to 250-fold. Parallel experiments were done to determine possible selective killing and it was ascertained that this would not account for the results obtained. The authors interpreted their results as a mutagenic effect of desiccation and of storage in the dried state, on the spores, as a consequence of injury to DNA by dehydration and the preservation of this injury in an altered state (ZAMENHOF et al., 1953; ZAMENHOF et al., 1956). Preliminary data from our laboratory also show that *Bacillus subtilis* $try^-$ $his^-$ spores stored under lyophilized conditions, have a higher proportion of re-

Table 12. *Azide-resistant mutants in spores of Bacillus subtilis strain 23 stored in dry state in vacuum*[a]. (From ZAMENHOF et al., 1968)

| Length of storage months | | 0 | | 7 | 91 |
|---|---|---|---|---|---|
| Conditions | | before drying | after drying | dry | dry |
| Starting wild | spore survival[c] | 1 | $6.7 \times 10^{-1}$ | $7.4 \times 10^{-2}$ | $8 \times 10^{-3}$ |
| | mutant frequency | $1.3 \times 10^{-6}$ | $2.5 \times 10^{-5}$ | $3.2 \times 10^{-4}$ | $2.5 \times 10^{-4}$ |
| Population mixture[b] | increase in mutant frequency[c] | 1 | 19 × | 250 × | 190 × |
| | spore survival[c] | 1 | $5.8 \times 10^{-1}$ | $7.1 \times 10^{-2}$ | |
| | mutants (%) | 56 | 57 | 23 | |

[a] Average of four determinations at 23° C. Washed spores of strain 23 in water ($10^9$/ml) were dried in glass bulbs over $P_2O_5$ and under high vacuum; they were then sealed and some samples were opened immediately to estimate effects of desiccation alone. Others were stored for 7 and 91 months. When pertinent spores from all the bulbs were resuspended in the original volume of saline 0.14M and tested for survival and resistant mutants.

[b] Mixture of Az-resistant and Az-sensitive parents was stored as in [a] to determine probable selection; "reconstruction" experiment.

[c] Comparative to first column ("before drying").

Table 13. *Biophysical characteristics of DNA isolated from spores and vegetative cells of Bacillus subtilis*[a]. (SPIZIZEN and EVANS: unpublished data)

| DNA characteristics | DNA from spores | | vegetative cells |
|---|---|---|---|
| | heavy | light | |
| Density | 1.703 g/cm$^3$ | 1.719 g/cm$^3$ | 1.703 g/cm$^3$ |
| Affinity for MAK | low | low | normal |
| Hyperchromicity on denaturation | unusually high | unusually high | normal |
| Biological activity | normal | little; only origin markers | normal |
| $T_m$ | normal | elevated ca. 5°C | normal |

[a] Analyses performed with purified DNA samples from spore preparations obtained by gradient centrifugation in renografin.

vertants than either spores stored in liquid suspension or vegetative forms of the same bacteria. Revertants seem to include both large and small colony formers (SERVIN-MASSIEU et al., 1970).

It is interesting to point out that HINTON (1965) has postulated that any damage suffered by organisms while in a cryptobiotic state cannot be repaired; thus, mechanical damage, damage by ultraviolet light or high energy radiation, are strictly cumulative.

Interesting differences between spore and vegetative form DNA have recently been found by SPIZIZEN and EVANS (unpublished). These differences, shown in Table 13 might be attributable to the elimination of water or to accompanying solute concentration taking place during the sporulation process. Perhaps these differences are also associated with the increased frequency of mutation observed in spores and in cryo-desiccated cultures of microorganisms.

## VII. Summary and Concluding Remarks

The preservation of microbial cultures by freeze-drying has been employed routinely in industry as well as in basic research under the assumption that this procedure does not alter any characteristics of the strain. However, a

Table 14. *Influence of inoculating dried spleen cells into Swiss mice*[a]. (From WEBB, 1965)

| | Number of animals | Animals with tumors (%) | Total no. of tumors[b] |
|---|---|---|---|
| Controls | 36 | 11 | 5 (3L, 2MC) |
| Non-dried cells | 24 | 8 | 4 (2L, 2MC) |
| Dried cells | 24 | 42 | 14 (7L, 7MC) |

[a] Pooled spleen cells from 1 year old mice were homogenized and centrifuged. Some cells were dried for 30 minutes. Animals received dried or undried cells intraperitoneally. After 12 months the animals were sacrified, autopsied and sections prepared for study.

[b] Figures in brackets indicate numbers of leukemias (L) and mammary carcinomas (MC).

number of reports suggest that the lyophilization of microbial cultures can introduce various types of genetic changes. These changes seem to be due to a direct effect of the process on the genotype of the treated cells and also to selective phenomena associated with differential survival of variant types in lyophilized populations. Among the many variables involved in the lyophilization process, the elimination of water, or dehydration, seems to be most responsible for genetic changes, since an elimination of water from the vicinity of DNA alters, in a significant way, the normal configuration of the hereditary macromolecule. In a natural system involving dehydration, namely sporulation, it has been determined that an increase in mutant proportions above that typical for hydrated, vegetative forms, can occur.

The problems discussed in this review require much more extensive investigation and it is hoped that this discussion may motivate further research along these lines. Eventually definite answers at the appropiate molecular levels must be obtained regarding the nature of genetic changes after removal of water from the cell, how the formation of spores provokes anomalies in the configuration of the genetic material it contains and the actual mechanism of this natural dehydration, how changes in hydration influence normal mitosis, etc. The reward for such efforts may conceivably even include solutions to problems with broad biological significance, like carcinogenesis, which, as indicated in Table 14, has been shown to have some relationship to simple, unesoteric molecules of water (WEBB, 1965).

*Acknowledgments*

The author wishes to express his gratitude to Drs. S. ZAMENHOF and W. BRAUN for valuable help in the preparation of this review and to Dr. J. SPIZIZEN and Dr. R. P. WILLIAMS for their permission to reproduce unpublished data. Also to the following authors and editors for permission to reproduce material employed: Applied Microbiology, American Society for Microbiology, Fig. 1 and Table 1; Dr. E. S. SHARPE, Dr. D. M. WHEATER and Journal of General Microbiology, Cambridge University Press, Table 2; Dr. F. W. PRIESTLEY and Journal of Comparative Pathology, Ed. Liverpool University Press, Table 3; Ciencia (Mex.), Ed. C. BOLIVAR, Fig. 1; Dr. D. H. BRAENDLE, Dr. S. M. MARTIN and University of Toronto Press, Tables 4 and 5; Dr. E. CABRERA, E. OLGUIN and Ciencia (Mex.), Ed. C. BOLIVAR, Tables 6 and 7; Dr. S. J. WEBB, Nature, Macmillan Journals Limited and Charles C. Thomas Publisher, Tables 8, 9 and 14 respectively; Dr. S. ZAMENHOF, Dr. H. H. EICHHORN, Dr. D. ROSENBAUM-OLIVER and Nature, Macmillan Journals Limited, Tables 11 and 12; Dr. C. AUERBACH and Zeitschrift für Vererbungslehre, Springer-Verlag, Fig. 3.

## References

ADAMS, M. H.: Bacteriophages, p. 418. New York: Interscience 1959.

APPLEMAN, M. D., SEARS, O. H.: Studies on lyophiled cultures: lyophile storage of cultures of *Rhizobium leguminosarum*. J. Bact. **52**, 209—211 (1946).

ARPAI, J.: Nonlethal freezing injury to metabolism and motility of *Ps. fluorescens* and *E. coli*. Appl. Microbiol. **10**, 297—301 (1962).

A.T.C.C.: Freeze drying procedure. Form No. 3. Rockville: Amer. Type Cult. Collec. 1964.

ATKIN, L., MOSES, W., GRAY, P. P.: The preservation of yeast cultures by lyophilization. J. Bact. **57**, 575—578 (1949).

AVERY, O. T., MACLEOD, C. M., MCCARTY, M.: Studies on the chemical nature of the substance inducing transformations of pneumococcal types. J. exp. Med. 79 137—158 (1944).

BALDWIN, R. L.: Kinetics of helix formation and slippage of the dAT copolymer. In: Molecular associations in biology, p. 145—162 (PULLMAN, B., ed.). New York: Acad. Press 1968.

BIRKHAUG, K.: Antigenic activity of fresh, frozen and dry BCG vaccine. Amer. Rev. Tuberc. **63**, 85—95 (1951).

BRADBURY, E. M., PRICE, W. C., WILKINSON, G. G.: Infrared studies of molecular configurations of DNA. J. molec. Biol. **3**, 301—317 (1961).
BRAENDLE, D. H.: Discussion I. In: Culture collections, perspectives and problems, p. 52—55 (MARTIN, S. M., ed.). Ottawa: U. Toronto Press 1963.
BRAUN, W.,: Variations in the genus *Brucella*. In: Brucellosis, p. 29. Washington: Amer. Assn. Advanc. 59. 1950.
— Bacterial genetics, p. 80—89. Philadelphia: Saunders 1965.
BUNTING, M. I.: The inheritance of color in bacteria, with special reference to *Serratia marcescens*. Cold Spr. Harb. Symp. quant. Biol. **11**, 25—32 (1946).
CABRERA-JUAREZ, E., OLGUIN, E. J.: Conservación por congelamiento del ácido desoxirribonucléico transformante de *Haemophilus influenzae*. Ciencia (Méx.) **26**, 65—68 (1968).
COLE, A.: Theoretical and experimental biophysics. New York: M. Dekker 1967.
DAMJANOVIC, V., RADULOVIC, D., VEN, S.: The effect of different gases on the stabilities of freeze dried suspensions of *Lactobacillus bifidus*. Proc. Symp. Surface Reactions in Freeze—dried Products, p. 1. Paris: Société de Chimie Industrielle 1969.
DAVIS, D. F., DULBECCO, R., GINSBERG, H., EISEN, H. N., WOOD, W. B.: Principles of microbiology and immunology, p. 169—182. New York: Harper 1968.
DEINSE, F.: Vaccination against tuberculosis with freeze-dried BCG vaccine. Amer. J. publ. Hlth **41**, 1209—1214 (1951).
DIMMICK, R. L., HECKLY, R. J. HOLLIS, D. P.: Free radical formation during storage of freeze dried *Serratia marcescens*. Nature (Lond.) **192**, 776—777 (1961).
ELEK, S. D.: *Staphylococcus aureus* and its relation to disease, p. 130. London: Livingston 1959.
FLOSDORF, E. W., KIMBALL, A. C.: Studies with *H. pertussis*. II. Maintenance of culture in phase I. J. Bact. **39**, 255—261 (1940).
FRIEDMAN, C. A., HENRY, B. S.: Bound water content of vegetative and spore forms of bacteria. J. Bact. **36**, 99—105 (1938).
GHEORGHIU, M., STURDZA, S.: L'Argon, tampon protecteur contre l'oxygene et l'humidite dans le conditionnement du BCG lyophilise. Proc. Symp. Surface Reactions in Freeze-dried Products, p. 1. Paris: Societe de Chimie Industrielle 1969.
GOODGAL, S. H., HERRIOTT, R. M.: Studies on transformations of *Haemophilus influenzae*. I. Competence. J. gen. Physiol. **44**, 1201—1227 (1961).
GORDON, D. E., CURNUTTE, B., LARK, K. G.: Water structure and the denaturation of DNA. J. molec. Biol. **13**, 571—585 (1965).
GORRILL, R. H., MCNEILL, E. M.: The effect of cold diluent on the viable count of *Pseudomonas pyoceanea*. J. gen. Microbiol. **22**, 437—442 (1960).
GREAVES, R. I. N.: Preservation of living cells by freeze drying. Ann. N.Y. Acad. Sci. **85**, 723—728 (1960).
GROSSBARD, E., HALL, D. M.: An investigation into the possible changes in the microbial population of soils stored at —15°. J. appl. Bact. **26**, vii—viii (1963).
GWINN, D. D., THORNE, C. B.: Transformation of *B. licheniformis*. J. Bact. **87**, 519—526 (1964).
HANAFUSA, N.: Denaturation of enzyme protein by freeze-thawing and freeze-drying. In: Freezing and drying of microorganisms, p. 117—130 (NEI, T., ed.). Baltimore: U. Park Press 1969.
HANSEN, I. A., NOSSAL, P. M.: Morphological and biochemical effects of freezing on yeast cells. Biochim. biophys. Acta (Amst.) **16**, 502—512 (1955).
HARRISON, A. P., PELCZAR, M. J.: Damage and survival of bacteria during freeze drying and during storage over a ten year period. J. gen. Microbiol. **30**, 395—400 (1963).

Hayes, W.: The Genetics of bacteria and their viruses, p. 179—200. New York: Wiley 1968.
Hearst, J. E., Vinograd, J.: The net hydration of DNA. Proc. nat. Acad. Sci. (Wash.) **47**, 825—829 (1961).
Heckly, R. J.: Preservation of bacteria by lyophilization. Advanc. appl. Microbiol. **3**, 1—76 (1961).
— Anderson, W. W., Rockenmacher, M.: Lyophilization of *Pasteurella pestis*. Appl. Microbiol. **6**, 255—261 (1958).
— Dimmick, R. L., Windle, J. J.: Free radical formation and survival of lyophilized microorganisms. J. Bact. **85**, 961—965 (1963).
Hinton, H. E.: Suspended animation and the origin of life. New Scient. **28**, 270—271 (1965).
— Reversible suspension of metabolism and the origin of life. Proc. roy. Soc. B **171**, 43—57 (1968).
Hinton, N. A., Maltman, J. R., Orr, J. H.: The effect of desiccation on the ability of *Staphylococcus pyogenes* to produce disease in mice. Amer. J. Hyg. **72**, 343—350 (1960).
Holton, G.: Scientific research and scholarship. Notes toward the design of proper scales. J. Amer. Acad. Arts Sci. **91**, 362—399 (1962).
Jacobson, B.: Hydration structure of DNA and its Physicochemical properties. Nature (Lond.) **172**, 666—667 (1953).
Januszewicz, I.: Effect of storage conditions on biochemical properties of *Leuconostoc mesenteroides*. Acta microbiol. pol. **6**, 367—376 (1957).
Jennens, M. G.: The effect of desiccation on antigenic structure. J. gen. Microbiol. **10**, 127—129 (1954).
Kabat, E. A., Mayer, M. M.: Experimental Immunochemistry, p. 114—116. Springfield: Ch. C. Thomas 1961.
Kaplan, R. W.: Auslösung von Farbsektor- und anderen Mutationen bei *Bacterium prodigiosum* durch monochromatisches Ultraviolet verschiedener Wellenlängen. Z. Naturforsch. **7b**, 291—304 (1952).
Lamanna, C., Mallette, F. M.: Basic bacteriology, p. 351. Baltimore: Williams & Wilkins 1965.
Lambin, S., German, A., Sigrist, W.: Influence de la lyophilisation sur le constituants et le pouvoir antigéniques d'*E. typhosa* et de *S. paratyphi* B. C. R. Soc. Biol. (Paris) **152**, 1650—1653 (1958).
Lea, C. H., Hannan, R. S.: Studies of the reaction between protein and reducing sugars in the dry state. Biochim. biophys. Acta (Amst.) **4**, 518—531 (1950).
Leach, R. H., Scott, W. J.: The influence of rehydration on the viability of dried microorganisms. J. gen. Microbiol. **21**, 295—307 (1959).
Leibo, S. P., Jones, R. F.: Freezing of the chromoprotein phycoerythrin from the red alga *Porphyridium cruentum*. Arch. Biochem. **106**, 78—88 (1964).
Levitt, J.: Status of the sulfhydryl hypothesis of freezing injury and resistance. In: Molecular mechanisms of temperature adaptation, p. 41—52 (Ladd-Prosser, C., ed.). Washington: Amer. Assn. Advanc. Sci. 1967.
Lincoln, R. E.: Control of stock culture preservation and inoculum build up in bacterial fermentation. J. biochem. microbiol. Technol. Engng **2**, 481—500 (1960).
Lion, M. B., Kirby-Smith, J. S., Randolph, M. L.: Electron spin resonance signals from lyophilized bacterial cells exposed to oxygen. Nature (Lond.) **192**, 34—36 (1961).
Lovelock, J. E.: Physical instability and thermal schock in red cells. Nature (Lond.) **173**, 659—661 (1954).
— The physical instability of human red blood cells. Biochem. J. **60**, 692—696 (1955).

LURIA, S. E., DELBRUCK, M.: Mutations of bacteria from virus sensitivity to virus resistance. Genetics **28**, 491—511 (1943).

MALTMAN, J. R., ORR, J. H., HINTON, N. A.: The effect of desiccation on *Staphylococcus pyogenes* with special reference to implications concerning virulence. Amer. J. Hyg. **72**, 335—342 (1960).

MARTIN, S. M.: Culture collections, perspectives and problems. Ottawa: U. Toronto Press 1963.

MAZUR, P.: Mechanisms of injury in frozen and frozen dried cells. In: Culture collections, perspectives and problems, p. 59—70 (MARTIN, S. M., ed.). Ottawa: U. Toronto Press 1963.

— Physical and chemical basis of injury in single celled microorganisms subjected to freezing and thawing. In: Cryobiology, p. 214—315 (MERYMAN, H. T., ed.). New York: Acad. Press 1966.

MERYMAN, H. T.: Principles of freeze drying. Ann. N.Y. Acad. Scie. **85**, 630—640 (1960).

— Cryobiology. New York: Acad. Press 1966.

MEYNELL, G. G.: The effect of sudden chilling on *Escherichia coli*. J. gen. Microbiol. **19**, 380—389 (1958).

MURRELL, W. G., SCOTT, W. J.: Heat resistance of bacterial spores at various water activities. Nature (Lond.) **179**, 481—482 (1957).

NEI, T.: Effects of freezing and freeze-drying on microorganisms. In: Recent research in freezing and drying, p. 78—86 (PARKES, A. S., and A. U. SMITH, ed.). Oxford: Blackwell 1960.

— Freezing and drying of microorganisms. Baltimore: U. Park Press 1969.

NEWCOMBE, H. B.: Radiation induced instabilities in *Streptomyces*. J. gen. Microbiol. **9**, 30—36 (1953).

NORTHROP, J. H., SLEPECKY, R. A.: Sporulation mutations induced by heat in *Bacillus subtilis*. Science **155**, 838—839 (1967).

NOVICK, A.: Mutagens and antimutagens. Brookhaven Symp. Biol. **8**, 201—215 (1956).

O'CONNELL, M. K., HUTNER, S. H., FROMENTIN, H., FRANK, O., BAKER, H.: Cryoprotectants for *Crithidia fasciculata* stored at —20° C, with notes on *Trypanosoma gambiense* and *T. conorhini*. J. Protozool. **15**, 719—724 (1968).

PLOUGH, H. H.: Spontaneous mutability in *Drosophila*. Cold Spr. Harb. Symp. quant. Biol. **9**, 127—137 (1941).

POLLARD, E. C.: The fine structure of the bacterial cell and the possibility of its artificial synthesis. Amer. Scient. **53**, 437—463 (1965).

POPOVSKY, V. G.: Physico-chemical and biological changes in fruits, berries and crushed juices during freeze-drying and storage. Proc. Symp. Surface Reactions in Freeze-dried products, p. 1. Paris: Societe de Chimie Industrielle 1969.

POSTGATE, J. R., HUNTER, J. R.: On the survival of frozen bacteria. J. gen. Microbiol. **26**, 367—378 (1961).

— — Metabolic injury in frozen bacteria. J. appl. Bact. **26**, 405—414 (1963).

PRIESTLEY, F. W.: Freeze drying of the organism of contagious bovine pleuropneumonia. J. comp. Path. **62**, 125—135 (1952).

PURKAYASTHA, M., HAMPTON, J., WILLIAMS, R. P.: An improved fixation method for the study of bacterial cell envelope by electron microscopy. Proc. Symp. Advancing Frontiers of Life Sciences, p. 60. New Dheli: Natl. Inst. Sci. India 1960—1961.

REY, L.: Conservation de la vie par le froid. Actualités scientifiques et industrielles, No 1279, p. 137—167. Paris: Hermann 1959.

— Traité de lyophilisation. Paris: Hermann 1960.

— Aspects theoriques et industriels de la lyophisation. Paris: Hermann 1964.

ROGERS, L. A.: The preparation of dried cultures. J. infect Dis. **14**, 100—123 (1914).

ROSS, K. F., BILLIG, E.: The water and solid content of living bacterial spores and vegetative cells as indicated by refractive index measurements. J. gen. Microbiol. **16**, 418—425 (1957).

SAITO, T., ARAI, K.: Slow freezing of carp muscle and inosinic acid formation. Nature (Lond.) **179**, 820—821 (1957).

SALT, R. W.: Application of nucleation theory to the freezing of supercooled insects. J. insect. Physiol. **2**, 178—188 (1958).

SCHRÖEDINGER, E.: What is life? p. 50—65. Garden City, N.Y.: Doubleday 1956.

SCOTT, W. J.: A mechanism causing death during storage of dried microorganisms. In: Recent research in freezing and drying, p. 188—202. PARKES, A. S., SMITH, A. U., eds. Oxford: Blackwell 1960.

SERRA, J. A.: Encyclopedia of plant physiology, p. 472—499. Berlin-Göttingen-Heidelberg: Springer 1955.

SERVIN-MASSIEU, M.: Spontaneous appearance of sectored colonies in *Staphylococcus aureus* cultures. J. Bact. **82**, 316—317 (1961).

— Mutagénesis por criodesecación de microorganismos. Ciencia (Méx.) **25**, 219—222 (1967).

— Inestabilización genética por liofilización de microorganismos. Rev. lat.-amer. Microbiol. **10**, 155 (1968).

— Inestabilización genética por liofilización de microorganismos: efecto diferencial de la congelación y de la desecación. Ciencia e Cultura (Brasil) **21**, 292 (1969).

— CRUZ-CAMARILLO, R.: Variants of *Serratia marcescens* induced by freeze drying. Appl. Microbiol. **18**, 689—691 (1969).

— FLORES, M., HERNANDEZ, M.: Manuscript inpreparation (1970).

— SANCHEZ-TORRES, L. E., PALLARES, F.: Gene unstabilization and mutation by freeze drying of bacteria. Bacteriol. Proc., p. 62. Detroit: Amer. Soc. Microbiol. 1968.

SHARPE, M. E., WHEATER, D. M.: The physiological and serological characters of freeze-dried Lactobacilli. J. gen. Microbiol. **12**, 513—518 (1955).

SHIKAMA, K.: The effect of freezing and thawing on the stability of double helix of DNA. Nature (Lond.) **207**, 529—530 (1965).

SIMMONS, E. G.: Fungus cultures: conservation and taxonomic responsibility. In: Culture collections, perspectives and problems, p. 100—110 (MARTIN, S. M., ed.). Ottawa: U. Toronto Press 1963.

SOUZU, H.: Localization of polyphosphate and polyphosphatase in yeast cells and damage to the protoplasmic membrane of the cell. Arch. Biochem. **120**, 344—351 (1967).

— Decomposition of polyphosphate in yeast cell by freeze-thawing. Arch. Biochem. **120**, 338—343 (1967a).

— ARAKI, T.: Injury on nucleic acids due to the freezing and thawing of yeast cells. Low Temp. Sci. B **20**, 69—79 (1962).

SPITKOVSKII, D. M., TSEITLIN, P. I., TONGUR, V. S.: Some effects caused by the two configurations of DNA. Biophysics (USSR) 1—11 (1960).

SPIZIZEN, J.: Transformation of biochemically deficient strains of *B. subtilis* by deoxyribonucleate. Proc. natl. Acad. Sci. (Wash.) **44**, 1072—1078 (1958).

— EVANS, A.: Unpublished data.

STAMP, T. C.: Preservation of bacteria by drying. J. gen. Microbiol. **1**, 251—265 (1947).

STEELE, K. J., ROSS, H. E.: Survival of freeze—dried bacterial cultures. J. appl. Bact. **26**, 370—375 (1963).

STEIN, C. D., MOTT, L. O., GATES, D. W.: Pathogenicity and lyophilization of *Pasteurella bubaliseptica*. Vet. Med. **44**, 336—339 (1949).

STILLMAN, E. G.: The preservation of Pneumococcus by freezing and drying. J. Bact. **42**, 689—693 (1941).

STRAKA, R. P., STOKES, J. L.: Metabolic injury to bacteria at low temperatures. J. Bact. **78**, 181—185 (159).
STRANGE, R. E.: Effect of magnesium on permeability control in chilled bacteria. Nature (Lond.) **203**, 1304—1305 (1964).
— DARK, F. A.: Effect of chilling on *Aerobacter aerogenes* in aqueous suspension. J. gen. Microbiol. **29**, 719—730 (1962).
SUBRAMANIAM, M. K., PRAHALADA-RAO, P. L.: Lyophilization and mutation in yeast. Experientia (Basel) **7**, 98—99 (1951).
— RANGANATHAN, B., KRISHNAMURTHY, S. N.: Reverse mutations in yeasts. Cellule **52**, 39—59 (1951).
SUTHERLAND, G. B. B. M., TSUBOI, M.: The infrared spectrum and molecular configuration of sodium deoxyribonucleate. Proc. roy. Soc. A **239**, 446—463 (1957).
SUZUE, G., TANAKA, S.: Carotenogenesis and resistance of *Micrococcus pyogenes* to tetracyclines. Science **129**, 1359—1360 (1959).
SWIFT, H. F.: A simple method for preserving bacterial cultures by freezing and drying. J. Bact. **33**, 411—421 (1937).
TAPPEL, A. L.: Effects of low temperatures and freezing on enzymes and enzyme systems. In: Cryobiology, p. 163—176 (MERYMAN, H. T., ed.). New York: Acad. Press 1966.
UMBREIT, W. W.: Modern microbiology, p. 203—204. San Francisco: Freeman 1962.
VELU, H., PIGOURY, L., COURTADE, R.: Comportement du Bacille morveux desséché sous le vide après congélation. C. R. Soc. Biol. (Paris) **136**, 775—776 (1942).
VOLZ, F. E., GORTNER, W. A.: The maintenance of *L. casei* and *L. arabinosus* cultures in the lyophilized state. Arch. Biochem. **17**, 141—148 (1948).
WASSERMAN, A. E., HOPKINS, W. J.: Effects of freeze drying on some enzyme systems of *Serratia marcescens*. Appl. Microbiol. **6**, 49—52 (1958).
WATSON, J. D., CRICK, F. H. C.: The structure of DNA. Cold Spr. Harb. Symp. quant. Biol. **18**, 123—131 (1953).
WEBB, S. J.: Bound water in biological integrity. Springfield: Ch. C. Thomas 1965.
— Mutation of bacterial cells by controlled desiccation. Nature (Lond.) **213**, 1137—1139 (1967).
— Effect of dehydration on bacterial recombination. Nature (Lond.) **217**, 1231—1234 (1968).
— Some effects of dehydration on the genetics of microorganisms. In: Freezing and drying of microorganisms, p. 153—168 (NEI, T., ed.). Baltimore: U. Park Press 1969.
— BHORJEE, J. S.: Infrared studies of DNA, water, and inositol associations. Canad. J. Biochem. **46**, 691—695 (1968).
— DUMASIA, M. D.: The induction of lambda prophages by controlled desiccation. Can. J. Microbiol. **13**, 33—43 (1967).
— — Bound water, inositol and the induction of lambda prophages by ultraviolet light. Canad. J. Microbiol. **13**, 303—312 (1967a).
— — An infrared study of the influence of growth media and myo-inositol on structural changes in DNA induced by dehydration and ultraviolet light. Canad. J. Microbiol. **14**, 841—852 (1968).
— TAI, C. C.: Lethal and mutagenic action of 3,200—4,000 Å light. Canad. J. Microbiol. **14**, 727—735 (1968).
WIKEN, T. O.: Fundamental aspects of cell preservation. In: Culture collections, perspectives and problems, p. 43—45 (MARTIN, S. M., ed.). Ottawa: U. Toronto Press 1963.
WILKINS, M. H. F.: Physical studies of the molecular structure of deoxyribose nucleic acid and nucleoprotein. Cold Spr. Harb. Symp. quant. Biol. **21**, 75—87 (1956).
WILLIAMS, R. E. O., RIPPON, J. E.: Bacteriophage typing of *Staphylococcus aureus*. J. Hyg. (Lond.) **50**, 320—353 (1952).

WITKIN, M. E.: Nuclear segregation and the delayed appearance of induced mutants in *Escherichia coli*. Cold Spr. Harb. Symp. quant. Biol. **16**, 357—372 (1951).

ZAMENHOF, S.: Effects of heating dry bacteria and spores on their phenotype and genotype. Proc. nat. Acad. Sci. (Wash.) **46**, 101—105 (1960).

— Gene unstabilization induced by heat and by nitrous acid. J. Bact. **81**, 111—117 (1961).

— Mutations. Amer. J. Med. **34**, 609—626 (1963).

— Study of microbial evolution through loss of biosynthetic functions: establishment of "defective" mutants. Nature (Lond.) **216**, 456—458 (1967).

— ALEXANDER, H. E., LEIDY, G.: Studies on the chemistry of the transforming activity. I. Resistance to physical and chemical factors. J. exp. Med. **98**, 373—397 (1953).

— EICHHORN, H. H.: Study of microbial evolution through loss of biosynthetic functions: establishment of "defective" mutants. Nature (Lond.) **216**, 456—458 (1967).

— — ROSENBAUM-OLIVER, D.: Study of mutability of stored spores of *Bacillus subtilis*. Nature (Lond.) **220**, 818—819 (1968).

— LEIDY, G., HAHN, E., ALEXANDER, H. E.: Inactivation and unstabilization of the transforming principle by mutagenic agents. J. Bact. **72**, 1—11 (1956).

# Author Index

Page numbers in *italics* refer to the references

# Subject Index

# Index to Volumes 40—53

Volume 46



Volume 47



Volume 48



Volume 49



Volume 50

Volume 51



Volume 52



Volume 53